U0133916

This Book Is a Tribute
to China's 30 Years of Reform
1978—2008

30 Reflections of China's 30 Years of Reform

1978–2008

FOREIGN LANGUAGES PRESS

First Edition 2008

ISBN 978-7-119-05439-1
© Foreign Languages Press, Beijing, China, 2008
Published by Foreign Languages Press
24 Baiwanzhuang Road, Beijing 100037, China
http://www.flp.com.cn
Distributed by China International Book Trading Corporation
35 Chegongzhuang Xilu, Beijing 100044, China
P.O. Box 399, Beijing, China
Printed in the People's Republic of China

Preface

Time really does fly — 30 years have whizzed by since 1978. The few big events which occurred that long ago, namely the matriculation of the first group of university students after the restoration of the national college entrance examination, the great discussion on practice as the sole criterion for testing truth, the reversal of the verdict on the 1976 Tian'anmen Incident, the fixing of farm output quotas for each household at Xiaogang Village of Fengyang County, Anhui Province, and especially the convening of the Third Plenary Session of the CPC's Eleventh Central Committee, all remain as fresh in my mind as if they had occurred only yesterday.

This has been the most vibrant period of Chinese history, during which the Chinese people have gradually expanded their living space, development opportunities and all kinds of rights and benefits. Thanks to the country's opening and reform policies, we have eliminated poverty, begun living a relatively comfortable life and, after decades or even centuries of self-isolation, China has gradually re-merged into the world community.

On the historical stage during that time, dramatic scenes have been acted out one after another which, resounding with pithy maxims and wise apothegms that would enlighten the benighted, have revealed so many people's prowess and sorrows, glories and dreams. Under the overall impact of reform, they have left their indelible marks on history.

History is made by people. The greatest change born of the opening and reform measures is a new system to help people realize their potential and fulfill their ambitions. In practice, everybody makes history and each human life reflects certain fragments of history. With this in mind, the Foreign Languages Press salutes the thirtieth anniversary of opening up and initial reform, along with the sacred Olympic Flame which is now illuminating the Chinese land, with *30 Reflections of China's 30 Years of Reform* in both the English and Chinese versions. This collection of articles unfolds the thirty years of history through the recounting of people's life experiences. I favor this type of narration because it shows our history more vividly and realistically.

After reading the first draft of this book, I have the following few thoughts:

First of all, the book upholds an extraordinary theme. During these past 30 years, China has undergone profound changes in all its areas of endeavor. This book is intended to trace the lives of ordinary Chinese people against the backdrop of the political as well as economic reforms and the sweeping changes in the social and economic fields in Chinese society. All the transformations are observed from the detailed aspects of personal lives, enabling readers to better understand the achievements and benefits of the reforms. Besides, such details will give them a deeper insight into the considerations that were the foundation for the reform initiatives.

Secondly, the book provides a broad vision. The 30 individuals selected for it include Yuan Geng, one of the earliest contributors to the country's measures of opening and reform; Professor Jiang Ping, who took pains to develop reform theories in cooperation with me; and private entrepreneurs who distinguished themselves in the transformative process and even today remain supportive towards the sound and rapid development of the national economy. Besides, most of the stories are about ordinary people from all walks of life and who live under vastly different circumstances. Readers with firsthand experience of this segment of Chinese history will almost certainly find their own more ambitious, soulful selves in those people.

Thirdly, the book is substantiated by a variety of angles. Whereas the story of one person alone can hardly cover all the social changes brought about during the past 30 years, the life stories of 30 people can reflect far more aspects of social-economic life in China. From the restoration of the national college entrance examination system, the return of educated youths from the countryside and the redressing of mishandled cases, through the implementation of the household contractual responsibility system in rural areas and the reform of the enterprise shareholding system, to the recent medical reform in rural areas, democratic management at the grassroots level, efforts to balance urban and rural development and to build the new socialist countryside, etc., the book tries to answer questions as to the impact of these strategic decisions and measures on ordinary people, not in the form of self-authoritative comments but with concrete, faithful records.

Fourthly, the book touches on a number of sensitive issues. It does not blindly sing the praises of past achievements, or turn away from social problems in the course of telling the grief, sorrow and pain of 30 individuals, including factory layoffs, the Three Gorges migrants, conflicts among different social groups, and the straitened conditions of peasant-turned industrial workers. The book is not for theoretical guidance, so there is no need to track down the causes of these problems or to actually find solutions. That leaves much leeway for further discussion or research.

Some foreign friends often ask me, "What, after all, has opening and reform brought to the Chinese people?" This book, I believe, will help to address those doubts to some extent.

Over recent years, the debate over whether to continue the reforms or not has surfaced from time to time. Some people think that, after thirty years, there is no reason to go much farther, and that any overhaul will mean the bankruptcy of some former policies. However, many who work in the fields of theoretical research, including myself, firmly maintain that the reforms have been very successful on the whole — the changes in the lives of the 30 individuals in the book have already proven this. Nevertheless, no policy can solve every problem in a single step. Take the main feature of the market economy, the enterprise shareholding system, for example. We used to argue a lot about whether to start that reform but found it somewhat premature to bring the system in line with established international practices, so we chose the "double-track operation system," whereby the incremental assets of state-owned enterprises would go public while their principal assets would remain temporarily out of circulation. Later, as the conditions matured and the disadvantages of the "double-track operation system" became all too apparent, we introduced a second shareholding reform to turn the "double tracks" into a "single track." This is what economists call "the split share structure reform."

It is said that the profound changes to Chinese society resulting from China's economic development during the past 30 years have exceeded those throughout the previous several thousand years. This remark is very interesting. Maybe it is improper to oversimplify the differences between periods of history, but the experiences acquired and accomplishments made in our successful transformation from the socialist planned economic system to the market economic system are evidently unparalleled. In fact, it is a pioneering undertaking in history. So long as we continue to be blessed with a broad vision, an open mindset and the entrepreneurial sprit of the past 30 years, China will amaze the world by creating even more miracles during the next 30 years. Despite uncertainties ahead, we will remain true to our beliefs.

Being part of all that has happened over all these years, I have hardly stopped recollecting, witnessing and pondering. There is, indeed, a reservoir of memorable, soul-stirring tales to share. Although I may not agree with all the authors of this assortment of articles, they arouse a genuine feeling of "returning" to life itself with all its details and "commonplace" occurrences. However, this sort of returning does not make us any less happy or proud to be living in such a great era. I believe that readers will feel the same — you will admit that the book is more than worth reading although you may not agree with some statements and perspectives in it.

Li Yining

July 18th, 2008

Guanghua School of Management, Peking University

Li Yining, economist, Dean and Professor of the Division of Social Sciences of Peking University, Dean Emeritus of Guanghua School of Management, Member of the Standing Committee of the 11th National Committee of the Chinese People's Political Consultative Conference (CPPCC), and Deputy Director of the Subcommittee of Economy of the CPPCC National Committee.

Contents

At the Starting Point of the Reform

| Yuan Geng | 1978: Acting Director of the Foreign Affairs Bureau in the Ministry of Communications (now Ministry of Transport), managing vice-chairman of the China Merchants' Steam Navigation Company. |
| | 2008: Retired |

"After the Spring Festival, your mother will go with me to Hong Kong to work. Take care of yourselves while we are absent, and be diligent in your studies and work," declared Yuan Geng at the reunion party on 1979's New Year's Eve. "For this good news, cheers!" That year Yuan Geng turned 62. In 1978, as the Acting Director of the Foreign Affairs Bureau in the Ministry of Communications (now Ministry of Transport), he had entertained the idea of retiring. However, just at that time, he received an order from the ministry to go to Hong Kong and take charge of the day-to-day work of the China Merchants' Steam Navigation Company.

CM (China Merchants) was at that time a subsidiary of the Ministry of Communications. Yuan Geng, who had for years been engaged in foreign affairs work, knew well the vicissitudes of CM during the last century: on December 23, 1872, Li Hongzhang (an important official in the late-Qing dynasty and a pioneer of the Westernization Movement) presented a memorial to the throne on starting the China Merchants' Steam Navigation Company, which later built up a group of modern industrial, transportation and financial corporations. In 1950, the staff of the Hong Kong CM, with 13 ships, staged an uprising against the Kuomintang by declaring allegiance to New China. When Yuan began to work in his position, however, its total capital was only 130 million yuan. He realized that the CM wouldn't survive without reforms. He hoped the central authorities would permit the CM to take full advantage of the favorable conditions in both Hong

Kong and the mainland, such as the capital and techniques of the former and lands and labor force of the latter. He expected the authorities to assign Shekou, part of Bao'an County in Guangdong Province that lies adjacent to the northwestern corner of Hong Kong, to the CM as the industrial area.

Yuan Geng's plan was sanctioned by the central authorities soon after the Spring Festival. But since there were so many things in China at that time waiting to be done and requiring money, the government couldn't provide much money for CM's reform and development. A leader said to Yuan Geng, "The Ministry of Communications can earn foreign currencies through synergies with Hong Kong. I hope you can buy ships and build ports on your own; the State won't give you any money. You must strive to survive and to develop. You are to earn foreign currencies and pay taxes, you need to discuss with the Customs, the Ministry of Finance and the banks; otherwise they will interfere if you are to follow different practices there."

Yuan and his wife left Beijing on the day that Deng Xiaoping returned from his visit to the United States. (Deng was the first Chinese leader to have visited the United States since the founding of the People's Republic of China in 1949.—tr.) Twenty minutes after Deng's special plane landed, they took off to Hong Kong via Guangzhou. The weather in south China was turning warm in early February. In the airport, Yuan took off his black worsted coat and handed it to his son, wishing to get down to work with a light pack, both literally and figuratively. He was the 29th leader of the CM, as well as the first managing director, chairman and Party Committee's secretary of the Shekou Industrial Zone in Shenzhen. Yuan was the one to fire the first cannon of reform in south China.

The first lesson Yuan Geng learned from capitalists was to make good use of money. The CM bought an office building, and the vendor required immediate payment. When Yuan wrote him a cheque for 20 million dollars, the vendor promptly drove to the bank to deposit it. The day was a Friday. The bank would be closed over the next two days. If he had failed to deposit this sum of money to the bank before 15:00 that day, he would have lost three days' interest. Yuan thought: people on the mainland wouldn't handle money that way. Later this lesson led to the famous motto: "Time is money," which echoed in the hearts of everyone involved in China's reforms.

A different event produced yet another famous motto: "Efficiency is life." The 600-meter-long Shun'an Dock project was the main one among the infrastructure projects managed by Yuan Geng. In 1979, at the beginning, the project followed the conventional equalitarian pay system, and the workers lacked enthusiasm. A worker only moved 20-30 trucks of earth in one day. The project progressed slowly. Yuan introduced incentives: the daily work quota for every worker was set at 55 truckloads of earth and whoever met the quota would receive a bonus of 0.02 yuan per truckload, and an excess bonus of 0.04 yuan for every truckload exceeding the quota.

These two occurrences inspired Yuan Geng, resulting in the slogan "Time is money; efficiency is life," which drew a strong response from the people. It was really courageous and insightful to put forward this type of slogan at that time when money was a taboo. The totally different concept of efficiency and values fitted in better with the rules of the market. Some scholars described the slogan as "a spring thunder that broke down barriers in people's minds." A foreign merchant investing in Shekou said that the value of this slogan was that it had a tremendous impact on the thinking of the Chinese people so that starting with Shekou, China was gradually transformed into a market-oriented country.

Yuan Geng also led the country in reforming the personnel

system. He abolished the system of life tenure, and publically invited applications for leading posts. The members of the board of directors were elected by the staff. Anyone who was supported by fewer than half of the staff members in the annual vote of confidence would be automatically dismissed. He also introduced into China the practice of public bidding for construction projects. All these methods, like an enlightenment campaign, caused profound changes in the outlook of the Chinese people .

The Shekou Industrial Zone was the test-tube baby of China's reforms. Large numbers of like-minded people gathered in Shekou, creating numerous miracles. Its efficiency was praised as the "Shenzhen speed." In just five years, the zone changed from an exit for people to flee to Hong Kong, to a place that attracted competent individuals and capital back. Seventy-four foreign-funded enterprises and joint ventures were established, among them 51 had gone into operation, with 14 already reporting profits. The average wage had become greater than that in Macao.

Deng Xiaoping, the chief architect of China's reform, came to Shekou for an inspection on the morning of January 26, 1984. Seeing the dock construction progress in Shekou, he was very happy and said, "You have built a dock, that is good! The system is fine." "The system is fine" were the first words of praise uttered by Deng during his inspection tour to Shekou and Shenzhen. As Deng was interested in this topic, Yuan gave him more details about Shekou, including facts about the economic system, organization, personnel system, wage system and housing supply reform. Finally, he said: "We don't know if we have been successful or have failed in these adventures that we have braved." Deng nodded and gave an affirmative reply.

Deng Xiaoping's approval of Shekou's motto "Time is money,

efficiency is life" excited Yuan Geng. On October 1 of the same year, a float with the slogan "Time is money, efficiency is life" appeared in the parade of the 35th National Day celebration that passed the Tian'anmen Square in the heart of Beijing. This slogan soon spread around China. In 1984, the reforms in Shekou cast the brightest aura in China. They were highly sanctioned by the state leaders. The reputations of the "special zone" of Shekou and of Yuan Geng reached a climax.

In September 1985, a finance company was established in the zone on the basis of the financial settlement center. As a non-banking financial enterprise, a finance company can both borrow money from banks and absorb deposits from enterprises. Its establishment further extended the fund-raising channels of the industrial zone, and enlarged the business scope of the settlement center, since the company could also provide services to enterprises outside the zone. On May 5, 1986, Yuan Geng wrote to the People's Bank, suggesting that the China Merchants Bank be founded. He felt that, with the experienced and advanced personnel management system of the Shekou Industrial Zone, the bank would definitely succeed. Obviously, this was a new experiment for Yuan Geng. The People's Bank held a meeting to discuss this proposal. The Governor of the People's Bank supported him and prevailed over all dissenting views and succeeded in getting the motion adopted. Developing from a simple settlement center of a company, the China Merchants Bank has now become one of the world's top 500 companies. In September 1990, Yuan Geng and the Shanghai authorities reached an agreement on the investment in projects such as developing the Waigaoqiao area. The Merchants Building was soon thereafter established in the financial center in Lujiazui. The Shanghai Branch of the China Merchants Bank was thus established.

China's reforms weren't all smooth sailing and there were major disputes along the way. Yuan Geng, always in the eye of a storm, was used to various kinds of criticisms. He was quite tolerant and generous about both the positive and negative assessments of him. In December 1992, Yuan Geng, who was then 72, received approval to retire. From then on, he faded out from the political scene and lived in seclusion. He was born in Shenzhen, reached the climax of his career in Shenzhen, and stayed on in Shenzhen after retirement. He spends his retirement by the sea beach in Shekou, a place which can never be separated from his name.

This revered old man says that in retirement, to stay in good health is his central task, adopting a generous attitude towards life. He goes to bed at 10 or 11 every night, wakes up at 5 or 6, and takes a nap at noon. He reads widely, from *The Communist Manifesto* by Marx and Engels to kung fu stories by Hong Kong writer Jin Yong and romances by Taiwan writer Qiong Yao. He always tries to buy and read new books, whether Chinese or foreign, that he has heard are interesting. He likes reading newspapers and subscribes to many newspapers and magazines or buy them at newsstands. He surfs the Internet, watches television, and chats with old friends. As is common for retired elders, his retirement life is peaceful and happy.

"I watch flowers blooming and withering in the yard, and I gaze at clouds folding and stretching in the far reaches of the sky." This couplet by a Ming-dynasty poet is often applied to someone who is indifferent whether in favor or out of favor. Now at the age of 90, Yuan Geng is a little weak, and he seldom goes out. He likes the trees at his front door, which were built not long after he arrived at Shekou. During his administration, he required people to plant trees in front of doors when building houses and did not allow them to cut the trees. An attractive environment attracts many entrepreneurs. They may establish their businesses in neighboring Dongguan or Huizhou, but they still prefer to live in Shekou. Yuan Geng likes the sea, which is not far from his window. He says that it is exciting to think of the sea, even though he can only stay in the house. His life belongs to Shekou.

The Idol of One Generation

Cui Jian	1978: Trumpet player for the Beijing Symphony Orchestra 2008: Musician

On one night in 1993 at the Capital Indoor Stadium, Cui Jian was performing a solo concert. With his long hair and wearing a T-shirt, Cui Jian was playing the guitar while hoarsely shouting out familiar rock rhythms on the single-cassette recorder. People who were present were so excited that they began to sweat profusely. Around Cui Jian were waving bare arms and swaying heads, and occasionally some drops of sweat fell onto his face and body...

This is a middle-aged man's deepest memory of that night. Before, posters of Cui Jian's concert could be found on campus overnight, though students could never afford the high price. Though he had no ticket, he nevertheless went to the Capital Indoor Stadium. And 20 minutes after the concert had started, he bought a ticket from a scalper for two yuan.

Cui Jian had been the idol of his generation for a rather long time.

At the age of 20, Cui Jian became a trumpet player for the Beijing Love & Peace Orchestra under the famous Beijing Symphony Orchestra. His career attributed to his growing environment. His father was a professional trumpet player, while his mother was a member of a dance troupe.

Later, Cui came to Beijing where he heard tapes that foreign tourists and students had brought to China and he fell in love with rock and roll. Inspired by the likes of Simon & Garfunkel and John Denver, Cui began to learn to play the guitar. Soon he could perform in public. In 1984, he formed Seven Ply Board (七合板) with six other professional musicians. They performed Western pop music at local hotels and bars. The band was the first of its kind in China. In the same year, he turned his creations into a disc named *Vagabond's Return*. Since he only wanted to show his musical style, he did not contribute lyrics and the quality of the recording was thoroughly substandard. He did not regard this disc to be his first real album.

By the mid-1980s, Western rock music had found its way into China. Famous bands included the Beatles, Rolling Stones, Talking Heads and the Police. Influenced by their music, Cui Jian tried his own hand at rock and roll. His earliest effort is a rock/rap piece entitled "It's Not That I Don't Understand." It's a pity that in 1980s' China, rock rap was so rare and unusual that few people could accept it.

In 1986, a concert was held at the Beijing Workers' Stadium in commemoration of World Peace Year, and the most popular Chinese singers all participated. The concert was sponsored by the China National Song & Dance Ensemble and Wang Kun was head of the ensemble. The person in charge of details was Hei Zi and he knew that Cui Jian had just written a song entitled "Nothing to My Name." Hei Zi talked to Wang Kun, saying: "Auntie, a singer named Cui Jian has written a song especially for this concert and I think it's very good, but the style is new." Wang Kun wanted to listen to it first. Then she pronounced , "Nice. Let him perform." At the concert, Cui Jian stepped onto the stage in peasant clothing and belted out his latest composition, "Nothing to My Name". As the song ended, a stunned audience erupted in a standing ovation. Soon, the song could be heard in many big and medium-sized cities. Its words were uninhibited and its music was bold and strange. Some people regarded it as an expression of "decadent capitalist culture" while others believed it had serious and thought-provoking content, such as resistance against the Cultural Revolution. For the young generation, Cui Jian's songs were truthful, could arouse passion in their hearts, and addressed such sensitive topics as freedom and sexuality; they were very excited to find an echo of this in Cui's rock music. Hence, Cui Jian, dressed in outdated and old clothes, became a great sign, a cultural symbol of enlightenment, ideas, reality and non-cooperation. At that time, it was in fashion for young people to bang out Cui Jian tunes on beat-up guitars on campus, at dormitories and in the streets.

In 1987, Cui had to leave the Beijing Symphony Orchestra because he had set the old revolutionary song "Nanniwan" to rock music and was thus forbidden from singing. He started to work with ADO, an innovative Beijing band including two foreign embassy employees: Hungarian bassist Kassai Balazs and Madagascan guitarist Eddie Randriamampionona. These and other foreign musicians introduced Beijing musicians to reggae, blues, and jazz, and their participation brought a rhythmic dynamism to Cui Jian's rough-hewn tunes. With ADO, in 1986, Cui Jian released what he considers to be his first real album, *Rock and Roll on the New Long March*. The phrase "New Long March" was often used in the official press at that time to refer to China's reform program. Cui hoped that his rock music could be promoted by jumping on the bandwagon. His songs in this album also mentioned "the great leader Chairman Mao" and showed that in his heart he shared the concept that art had a function of helping the cultivation of a sound social conduct and had walked out of the shadow of "Nanniwan" onto a correct road. However, rock music, in contrast to China's folk and traditional music types and pop music genres, was undoubtedly branded with criticism from its birth. But Cui Jian's criticism was different from that of others. He could make people reflect on their attitudes to life instead of blaming others. All the songs in this album were classical and could be shocking. As a comment of that time went: "The release of the *Rock and Roll on the New Long March* ended the state of wasteland in China's pop music. Cui Jian's rock music represents a kind of belief, a kind of persistence and a kind of true self; it's true music."

While he was making a name for himself at home, Cui Jian was also beginning to receive recognition from abroad. In 1988 he performed "Nothing to My Name" in a special live world-wide broadcast for the 1988 Olympic Games in Seoul. In 1989 he was invited to the first Asian Popular Music Awards held at the Royal Albert Hall in London, the United Kingdom, and then to the "Printemps de Bourges" Festival held in Paris, France. After returning to China, Cui began his first rock tour entitled the "New Long March." His performance decorated Beijing for the coming Asian Games and lightened up the rock circles on the mainland of China.

In 1991, Cui Jian released his second album *Solution* which consisted of songs written before 1990, and continued to experiment with his sound. With his fame established, this album was popular in China. To his surprise, his album was also popular in Japan. In 1992, when MTV from Hong Kong came into vogue on the mainland of China, one track from Solution entitled "Wild in the Snow" was made into an MTV video which received an MTV International Viewers' Choice Award and became an instant hit throughout Japan. Cui Jian gave his first performance in Tokyo. In 1994, Cui Jian released his third album, *Eggs Under the Red Flag* which won acclaim both at home and abroad. Cui undertook a four-city tour of Japan to promote this album. He was as popular on this tour as he had been eight years before on the mainland of China. In 1995, he made his American debut and later undertook a tour which received positive comments from all the local media.

Cui Jian continued to look for fresh elements for his music. In 1998, he released his fourth album *The Power of the Powerless*. A work of digital avant-garde rock, this record is a marked departure from his previous efforts. Harnessing the semantic density of rap music, Cui Jian sketched China's changing social and economic landscape at the end of the 20th century. But this attempt failed and people paying close attention to him said "Cui Jian is old." Cui Jin did not acknowledge that. He worked for five years with great concentration and in March 2005, he released his fifth album *Show You Colors*. He wanted to use this album to show "some colors" or to "teach a lesson" to his audience who had been disappointed with him. However, he only received comments such as: "This album replaces his past pungent songs that could go right into your heart with formalistic lousy clichés." His fans, who had been hungry over a decade ago, are now no longer rebellious but pursue a stable life and are used to enjoying slow music after a busy workday. While today's young people, also children of Cui's fans, now have bars, discos, football matches and electronic games in their lives. Regarding music, they prefer Jay Chou and Supergirls. Perhaps, just as some critic has said, in the current China with a diversified culture and more and more stable life, people are no longer as rebellious as they were 30 years ago, and don't need to vent their depression through shouting.

On January 5, 2008, Cui Jian, the 47-year-old "Father of Chinese Rock," held a concert which was a review of his classics in over 20 years at the Beijing Workers' Stadium, where he had shouted "Nothing to My Name." Since at the same time and at the same place, there were several concerts of Cai Qin, Shin, Yu Quan and Han Hong, Cui Jian made his ticket the cheapest. Ten days before the concert, only one fourth of the tickets had been sold. Thanks to promotion, finally nearly 10,000 people came to his concert.

People who have been to his previous concerts sigh, Cui Jian is really old. Just as he said in the song "It's Not That I Don't Understand," the world changes so fast and you can never catch up with the times. However, Cui Jian's pursuance has left a deep impression on one generation. In memory, they can still hear Cui's song of encouragement: "I want to leave, /I want to exist, /I want to start all over again after death..."

A Teacher's 30 Years

Wang Jiacong | 1978: A primary school teacher in Lijiabao in
Baqiao District, Xi'an City
2008: Retired

Wang Jiacong, 76 years old, taught in Baqiao throughout his working years. He started with a salary of RMB 48 yuan per month in 1952. The sum later rose to 62 yuan, and remained unchanged until the end of the 1970s. Now he receives a monthly pension of 1,800 yuan, 30 times more than what he made before China's opening-up and reform.

In 1952 the 20-something Wang took a teaching post with a primary school in Lijiabao, Baqiao District. The starting salary of 48 yuan a month was barely enough for a family of three to eke out a living. After the salary reform in 1956, his payment leaped to 62 yuan, a relatively high rate in China at the time. From that time on, he began to put aside 20 yuan every month. His savings didn't grow to a big amount, as he had to give a sum to both his parents and his wife's parents at every Chinese New Year. His salary saw no big change in the following two decades. In the 1970s the average per capita disposable income was around 300 yuan in Xi'an, capital of Shaanxi Province. From the beginning of the 1980s, Wang saw steady growth in his income. By 1989 he was making 500 to 600 yuan per month. But prices had also increased. Besides, his family had expanded to five, and he still had to support his parents and parents-in-law. He nevertheless managed to squeeze out 100 to 200 yuan for his savings each month. When Mr. Wang retired in 1993, he received a pension of more than 1,000 yuan. After all his three children got married, Wang and his wife began to see an annual surplus of 4,000 to 5,000 yuan in their family budget.

This type of financial improvement can actually be seen in every family in the city. In 2001, the average per capita disposable income of people from Xi'an reached 6,705 yuan. In 2007, their per capita monthly income exceeded 1,000 yuan for the first time in history, adding up to 12,662 yuan per head, 30 times that in 1978.

Seeing many people around him buying stocks and funds, Mr. Wang also jumped on the bandwagon.

Clothing: From one set of clothes for all seasons to multiple sets of clothing for each season

"I never worry about the cold weather in winter, for my children have bought me several warm coats of different thicknesses," bragged Wang. When the temperature dropped to record lows after a rarely seen snowstorm in the city, Wang still went out to buy groceries everyday, comfortably clad in a new down jacket from his son. This is in sharp contrast with the past. Wang remembers that when he was young, he wore the same one or two items of clothing throughout the year, and children could get new clothes only at the Spring Festival. Once he bought a piece of cloth for his daughter, and explicitly told the tailor to make the shirt a bigger size, so that the girl wouldn't outgrow it too quickly . The tailor however made it an exact fit. The couple was peeved over the incident for several days.

Owing to the short supply of commodities, the government implemented a rationing system in the 1960s and 1970s. Many consumer goods had to be purchased with ration coupons. And for those goods in particular shortage, a special token was required. Every citizen in the city was entitled to a token of ten credits every month. And a varied amount of credits was deducted based on the purchase of different goods. When Wang's daughter got married, he bought a piece of woolen material, which was in fashion at the time, from Shanghai with someone's help, which cost him three tokens. This meant that he didn't have enough credits to buy other necessities that month.

Since the early 1980s, garments of more varieties and amounts appeared in the market. Some were made of materials that had never been heard of before. By 2007, the expenditure on clothing by the people of Xi'an rose from 73.7 yuan in 1980 to 950.5 yuan per head. Once people can afford more clothes, they become picky about the design and brand name.

Now Mr. Wang gets a couple of new items of clothing at every turn of the season, some of established brands. "Getting clothes is no difficulty nowadays. I can buy them whenever I need to."

Eating: From eating for mere subsistence to eating for culture and health

Before the early 1980s, most Chinese yearned for nothing more than to be able to bring enough food to the table. Now they stress the nutritional and caloric content of their diets.

When he arranged for the Spring Festival dinner, Mr. Wang opted for seafood and vegetables instead of red meat or poultry. "In the past red meat and poultry were precious. They are not nowadays. My children always urge me to have a balanced diet: stay away from fat, more yogurt, no more than an egg a day, fruits in the morning and evening, etc."

In the 1950s, Wang's family had meat only once a week. In the 60s and 70s one couldn't buy meat without ration coupons, distributed to households according to their size, even if one could afford it. Every family had to calculate its grain consumption, or it couldn't make ends meet for the month. After China adopted the opening-up and reform policy, the market flourished, and rationing was eventually abolished. "Grain was the last rationed item. A coupon for 500 grams of grain could be traded for ten eggs at first, and gradually depreciated to 5 kilograms for one egg. I, however, held tight to my grain coupons, and never made such deals. They are now worthless scraps, but tell my grandchildren a piece of our history."

Today most families won't collect big stocks of food ahead of holidays as they did before, for the market has abundant supplies throughout the year. Sometimes the Wangs eat out for family gatherings, so that the couple is spared the chores of cooking and cleaning.

Housing: He moved five times, each home bigger than the previous one

It is the desire of everyone to own a home. The average per capita living space in Xi'an was 9.9 square meters in 1998, a 90.4 percent growth over that in 1980. It rose to 23.6 square meters in 2007.

The Wangs have had five homes in the past years. The first was a rental home in a rural community adjacent to Mrs. Wang's working unit. It sheltered the family for 20 years. Despite the low rent, merely 4 yuan a month, living there was no pleasant experience. "It was freezing cold in winter. What's more, we were always scared by the whining of wolves on our way home in the evening."

In the early 1970s the family moved to a room in a residential building of Mrs. Wang's working unit. Though they had to share it with another family, and all occupants of the floor shared one toilet, they were delighted about moving into a "public-funded house."

At the end of 1970s the Wangs moved into another room, one with a kitchen. "The days were over when we had to jostle with another family when cooking meals. That marked a new stage in our lives." In the 1980s the Wangs eventually moved into a remodeled two-bedroom apartment with both kitchen and toilet. "That's all one aspired to at that time." Wang said.

In the 1990s Mr. Wang was assigned a new apartment with two bedrooms and one sitting room. He therefore gave the old apartment to his newly married son. The new home was not big, but had nice surroundings. Now he considers buying a bigger apartment nearby as an investment.

Transport: From bicycle to airplane

Mr. Wang bought his first bike in the 1950s, a size-28 Forever man's bicycle, which cost more than 100 yuan, a big sum at the time. In those days a bike was deemed one of the required items to make a decent wedding, together with a sewing machine, a watch and a radio. Wang rode his bike to work and back home every day, to meetings out of his school, and to his hometown in Fuping County.

Later a bus route extended to Fuping. But Mr. Wang had to change two buses on the way, and then walk for several miles before getting to his parents' home. "If I made the trip with a lot of baggage, I would be affected for several days afterwards." A direct bus to the county was launched in the 1990s, making Wang's visit to his parents much easier. Now there are a good number of buses and taxis in the street. And most people can afford an air flight for trips out of the province. "We never thought that someday we could have such a developed three-dimensional transportation network," Wang exclaimed.

Home Appliances: From black-and-white TV to LCD

When Mr. Wang bought an LCD TV for 10,000 yuan two years ago, he didn't feel as excited as when he got his first black-and-white set in the 1970s. Seeing his neighbors one by one taking home the magic box, Wang made up his mind to get one himself after a month's hesitation. The set cost him approximately 300 yuan, several years of family savings at the time. "My family and some neighbors gathered in front of the TV in my home every night. With two or three channels, it still brought us tremendous joy."

The demand for electric home appliances surged rapidly following the start of China's opening-up and reforms. When Wang's first TV broke down in the 1980s, he replaced it with an 18-inch color TV of a domestic brand. During the years more and more electric products found their way into his home, including a washing machine and a refrigerator.

In the 1990s, a microwave oven, range hood and air-conditioner appeared in Wang's home. The latest addition is a computer, with which Mr. Wang follows the ups and downs in the stock market every day. These modern items have changed the way of life for his family.

A Veteran Construction Engineer Who Helps Change Beijing

| Guan Geng | 1978: Engineer of Beijing Construction and Engineering Bureau |
| | 2008: Retired |

Taking part in the construction project of the National Grand Theater (now renamed National Center for the Performing Arts) was the highlight of Guan Geng's career as an engineer. Starting from 2000, he had been a supervisor on the site of the theater. Since the National Grand Theater was a key project of China in the new century, the Chinese government attached great importance to it and thus construction workers enjoyed very good conditions. The workers were migrants from around China. Before they started to work on the site, the contractor company had prepared good logistics for them. Since they lived in dormitories uniformly built in the suburbs of Beijing, far away from the site in the city center, the contractor sent buses daily to shuttle them between the dormitories and the site. Their meals consisted of Chinese fast food centrally delivered, including both meat and vegetable dishes. As a supervisor, he saw the quality of this project was very good. Workers proved themselves eqaul to their task which required exact technological standards. To make the height lower than that of the Great Hall of the People, the parliament building standing on the other side of the street, the designers limited the height by making the revolving stage 33 meters deep underground. When excavating the earth, truck drivers needed to drive down the slope to the bottom, which demanded a high degree of proficiency. A dozen trucks came and went in order, making the scene spectacular. Guan Geng said that the scene had left a deep impression on him. No big errors occurred, showing that the workers were skillful and well organized. On the site, workers worked with enthusiasm. Guan estimated that for a period of time, there were over 1,000 people and a hundred trucks working on the project. Viewed from a distance, the theater looks like a huge eggshell covered with titanium steel. Workers were tied with manrope and paved one piece after another along the slope. They did a careful job, while the contractor company also did well in terms of security control. No big accident occurred for such a large eggshell. This was no easy accomplishment.

It's a pity that Guan did not see the completion of the project. Since 2004, as a result of cerebral thrombosis, he has not been able to move his left leg easily. He needs to rest. Hence, it has become part of his life to recall his past.

On the top of a bed stand in Guan Geng's bedroom, there is a yellowed map of 1950 Beijing mounted in a plank frame with glass. When he has nothing to do, Guan likes to put on his glasses and carefully look for places he has been to, though changes have taken place in these places. He grew up at a siheyuan (a compound with houses around a courtyard) in Beijing and then worked in the building industry for nearly 40 years, and thus witnessed more of the changes that took place in Beijing than ordinary people. Sometimes when he saw places changing gradually as a result of his work, he was more joyful than nostalgic.

The reform and opening-up starting in 1978 was a key turning point in Guan Geng's life. As an engineer at the Beijing Construction and Engineering Bureau, he started to take charge of some large buildings. After the Cultural Revolution, a lot of projects needed to be done in terms of city building in Beijing. However, Beijing had a fairly weak foundation in terms of building technologies, with only several stacker cranes; thus it was difficult to build tall buildings. For the refrigeratory project in the southwest suburbs of Beijing, Guan

Geng and his colleagues adopted the technologies of slipping and slab-lifting, which were advanced at that time, and thus solved the problem of having no stacker crane. A few years later, building technology had developed rapidly, and several big construction companies set up branches in Beijing in succession. Consequently, large buildings sprang up like mushrooms.

Guan Geng became busier as he took charge of more and more construction projects. He had to work from morning to evening at every site. Working overtime was standard. He could only have one day of rest every two weeks. Sometimes he was too busy to have meals, and he found himself exhausted after work. However, he seldom felt tired, because he was supported by a kind of work enthusiasm. Construction workers worked very hard; they were mostly employees with long years of service and proud of their trade. Construction companies have an annual performance assessment and regular exams on some practical problems occurring in construction. Encouraged by honor, workers worked in a very careful manner. Sometimes, when problems relating to quality control occurred, the workers would come to work even at mealtimes.

With an increasing workload, Guan saw his salary rising and had some spare money. In the early 1980s, it was in vogue to drink beer with meals. Since beer was in short supply, beer breweries produced beer in bulk. Beer in bulk was not as cool and full of bubbles as today's beer on draft; it was transported in big tank trucks to shops and then pumped into containers, and thus was warm and had a different taste every time you drank it. Moreover, people had to queue up early; otherwise they would be unable to buy it. In the afternoon, when Guan Geng figured out that beer was coming, he would, as would his neighbors, carry his own container and queue up to buy beer. Once he was home, he cooled the beer with water and drank it with his meals.

Some people who were very fond of beer drank it in the street as soon as they purchased it. Guan could not have known that less than ten years later, ordinary people would be able to go to bars to drink beer, or buy several boxes of canned beer and drink frozen beer at home. During that period, life improved at a speed that people could hardly believe.

At that time, the garment industry still followed the same old beaten track and had not yet caught up with people's changing taste. Hence, many peopled tried to make their own clothes. With knowledge of design drawing, Guan Geng boldly started with simple costume designs. After working on it for a while, he knew that it was easier than drawing engineering designs. Later, Guan Geng bought a sewing machine. He drew the design, and his wife sewed the clothes. They made shirts, coats, skirts and clothes for their daughter, all of which were praised by their neighbors.

In the 1980s, with his increasing income, Guan Geng bought many electronic appliances. He obtained a coupon for a television, and then bought a 12-inch Sanyo television set at the Qianmen Electronic Appliance Shop. From then on, his daughter would not have to go to his neighbor's to watch TV. Every day after dinner, the whole family gathered in the courtyard and watched TV, while neighbors who had no TV also came with stools. Short ones sat in front of tall ones — it was very noisy, just like an open-air theatre. At that time, refrigerators were getting popular but could only be bought with coupons. Buying his first refrigerator caused Guan Geng a lot of trouble. He got the coupon from the site of the Hardware and Electronic Appliance Factory. Since he had to pick up the refrigerator by himself from the warehouse far away from home, Guan borrowed a tricycle, and went to the warehouse with his wife and transported the refrigerator home. It was a single-door Wanbao Brand refrigerator. When they drank their first mouthful of cooled soda, the whole family was very happy.

In the 1990s, to maintain a good ecological environment in Beijing, a new strategy of urban development was proposed, which was to adjust and renovate central areas instead of constructing new buildings as had previously been done. In 1992, when Zhou Kaixuan, who worked with the Oriental Overseas Corporation of Hong Kong, got to know the reconstruction plan of the East Chang'an Street and Wangfujing Shopping Area in Beijing, she proposed the building of a new Oriental Plaza and persuaded Li Ka Shing, a leading businessman in Hong Kong, to invest US$2 billion into this project. From 1993 to 2000, Guan Geng was the Assistant Chief Engineer for the Oriental Plaza Project under the Beijing Construction Corporation. In the seven years from inhabitant resettlement to completion of the project, Guan Geng had been working day and night at the site with the workers. Finally, they built a modern and multi-functional building on a plot of 760,000 sq m. The Oriental Plaza was the largest single building in Beijing. Now, every time he stands in front of the radiant Oriental Plaza, he will remember the scene in which his elder brother and he sat on a step and counted trams. Although it is totally different now, he is still excited to think that he himself also made contributions.

Guan Geng cherishes fond memories from his past and always wants to recount his happy hours. When he was a supervisor of the National Grand Theater project, for a while he was not so busy since the construction had been halted to wait for the design proposal. Hence he went one or two hours earlier every day and drew pictures depicting his childhood, family, studies, love affair and marriage... Not giving much thought to his drawings, Guan drew as he wished in his spare time and was able to fill up three notebooks in two or three months. Later, the editors of China Youth Press saw his drawings and judged them to be the best way to reflect the changes in Beijing in the previous century from the perspective of an ordinary resident. In 2007, Guan's drawing book *My Previous Century — Drawing Book on the Private Life of a Beijing Civilian* was published. The book includes many pure folk customs of old Beijing and the most precious memories of Guan Geng.

For Guan Geng, what has made him the happiest in recent years is that his home has become bigger and bigger. He has moved four times, from siheyuan to an 18-square-meter one-room home, then to a 30-square-meter two-room home, and finally to his present nearly 90-square-meter three-room home. By the end of 2008, his family will move into a new, bigger home. Although their home becomes more and more spacious, the family preserves the same atmosphere of a big family as when they lived in a siheyuan. His family is now made up of four generations, from his old mother-in-law to his little grandson, all enjoying the happy times. To catch up with the times, Guan has his home equipped with five televisions, one for each room. And, to watch various kinds of video tapes and CDs, he has also bought a video recorder, VCD and DVD. With the upgrading of technologies, three computers have appeared at home. Moreover, a laptop, digital camera, MP3 with radio, and other digital appliances have appeared at his home in this information and networking era, just as they have in all ordinary households in Beijing. Guan Geng is very satisfied with his present state: he reads books at home, raises several small tortoises as pets, has a walk with his wife in the park on sunny days; his life is simple, comfortable, and also happy.

Shouting out for Rights

Jiang Ping | 1978: Back to Beijing College of Political Science and Law
2008: Tenured Professor of China University of Political Science and Law, tutor of Ph. D. students majoring in civil and commercial law, Vice-Chairman of China Law Society

In 1978, the Beijing College of Political Science and Law was re-established, and Jiang Ping came back to teach.

After his return to teaching, Jiang Ping came to a profound understanding of how much importance the leadership attached to law. Due to the Cultural Revolution, China's law education and study had stopped for more than 10 years. Some old professors who had taught prior to the Cultural Revolution had reached the age of retirement, and works of law had been almost completely destroyed during the Cultural Revolution. Jiang Ping deeply cherished his profession as a teacher, hoping to impart his knowledge and the concepts of law to young people and to develop them to be pillars of the nation in restoring the rule by law. To cultivate students with the spirit of the law, teachers had to impart a complete system of ideas about law, but the education on private law was prohibited at that time. The shadow of the Cultural Revolution had not been completely dispelled, and people were still afraid of talking about anything "private" in such a social environment. Jiang Ping was keenly aware that Chinese society lacked the spirit of private law. He made a proposal to start two courses on Western countries: one was Civil and Commercial Law and the other was Roman Law. Students found the two courses novel and were curious about them, while at the same time they felt uneasy or even found them improper. In the past, the course on foreign civil and commercial law was all about criticism, and even the name of the course had "criticism" in it. However, Jiang Ping introduced it to Chinese students as a formal course. The spirit of Roman Law is that of private law. At that time it was regarded as defiance of state power to advocate "private law." But Jiang Ping expected his students to understand that only when private rights are guaranteed, can the state achieve the rule of law.

In 1985, Jiang Ping who had just become vice-president of China University of Political Science and Law accepted the task of drafting the General Provisions of the Civil Law, as a member of a four-member expert group. In the process of drafting the Provisions, Jiang widely advocated the significance and functions of the General Provisions of the Civil Law in many forms, and explained many difficult problems about legal concepts. His research and explanations enabled the drafting, approval and implementation of the General Provisions of the Civil Law to proceed smoothly. In August, the expert group finished its first draft. Later, they held repeated discussions and completed the revised edition, which was adopted on April 12, 1986. The General Provisions of the Civil Law has been regarded as China's Declaration of Rights, as it endows the Chinese people with the rights of private law. Owing to his good work, Jiang Ping has been respectfully addressed as "Mister of Civil Law" in legal circles.

For Jiang Ping, the legislation of the General Provisions of the Civil Law is not the accomplishment of which he is most proud. The Civil Law appeared as early as during the Republic of China (1912-1949), and its prototypes can be traced back to the late Qing Dynasty (1644-1911). Moreover, each country has its own Civil Law. The real breakthrough he made was in the enactment of the Administrative Procedural Law.

In 1983, Jiang Ping led a team in investigating the legal systems of Belgium and West Germany. To his surprise, both countries had special courts for administrative proceedings, and most of the plaintiffs were citizens while the defendant for all cases was the government. What surprised him more was that the plaintiffs won 30% of the cases. Before the tour, he had been thinking over a question: 5,000 years of feudal rule had made the Chinese people accustomed to obedience, in an effort to bring their words and deeds in conformity with the framework of public rights. He had been pondering the following question: was it possible to mark the boundary of public rights, to take public rights under supervision and provide timely and effective protection for private rights when infringed upon by public rights? In 1987, the first anniversary of the implementation of the General Provisions of the Civil Law, a vice-chairman of the National People's Congress (NPC) Standing Committee suggested that Jiang Ping organize a team on administrative legislation, and Jiang Ping agreed. Two years later, the Administrative Procedural Law that he was in charge of was adopted. On the day when it was adopted, Jiang Ping, now a member of the NPC Standing Committee and deputy director of the Commission of Legislative Affairs of the NPC Standing Committee, was extremely excited, because from then on China had a system under which civilians could take officials or governments to court, and China, as a socialist country, also had its own Administrative Procedural Law.

The Administrative Procedural Law is perhaps the law that offers the most legislative guarantees for Chinese civilians in suing administrative departments. In the last two decades, courts at various levels in China handled over 50,000 cases involving civilians suing officials or government departments each year. And the scope of these cases have expanded, from several fields such as resettlement and public order management to the current categories of intellectual property rights, land and mineral resources, city planning, education, state-owned capital, networks and more, covering almost all fields under administrative management and all administrative departments.

Later, Jiang Ping took charge of and participated in the legislation of other important laws of China. He was head of two respective expert teams for drafting the Contract Law which was released in 1999 and the Trust Law which was released in 2001, and participated in the legislation of the Law on State Compensation, Company Law, Securities Law, Negotiable Instruments Law, Partnership Enterprise Law, Law on Proprietorship Enterprises and others. Although he is now over 70 years old, Jiang Ping still takes charge of the expert team for drafting the Property Law and Civil Code. Owing to these laws, those learning law regard Jiang Ping as an outstanding individual, while more other Chinese people enjoy the good changes in their lives brought about by Jiang Ping although they don't know him.

In 1988, Professor Jiang Ping was promoted to the post of president of China University of Political Science and Law. From that year on, he insisted on giving the first lesson to freshmen. Standing before innocent young people around 20, he talked little about "Chinese characteristics" but rather emphasized the "universality" of law. He spared no effort in advocating "rights" and in his eyes, the rights of the state and the rights of the public were equally sacred. He helped to connect China's law education with that of the world and integrate it into the legal system of mankind. Today, the president and students of the university are different, but China University of Political Science and Law, with this distinctive characteristic unique in China's law education, is beyond dispute the most prestigious university of law in China.

As president, Jiang Ping was respected by students for his knowledge; but he was also warm and friendly to the students. In the classroom, he was always in the back row sitting in on the class; in the dining hall, he ate his meals sitting alongside the students; in the dormitory, he talked with students until lights went out... Students regarded him as a friend. To train students with the real qualities of the law, Jiang Ping at all times gave priority to teaching over all other university work. He kept many good teachers and helped departments concerned to introduce good teachers, and always offered the most important help to teachers at the most difficult moments. Jiang Ping implemented two ideological pivots for the rule of law — democracy and freedom — into the operation of the university and thus was acknowledged as "president of democracy."

"I'm not a 'jurist', I'm only an educationist." This is what Jiang Ping says modestly, and how he describes his later years. In the 1990s, Jiang Ping devoted himself to organizing the translation of Western legal works. As early as in 1988 when he visited Italy, Jiang Ping reached an agreement with the Roman Law Expansion and Research Team under the Italian National Scientific Research Committee, and decided to translate original documents of Roman Law. It was a huge project, and it will take 20 years to translate all the 5 million words in over 20 volumes. It was almost beyond imagination at that time. However, with his foresight and sagacity, Jiang Ping was able to provide firsthand historical documents and materials to China's law education and research on the Roman Law. And he also organized the translation of a series — "Library of Foreign Laws" with financial aid from the Ford Foundation, which included more than 30 laws of over 10 million words.

Now a tenured professor of China University of Political Science and Law, the nearly 80-year-old Jiang Ping is still instructing some Ph.D. candidates. He, Wu Jinglian and several other respected leading scholars of different academic fields often gather together and comment on the hottest issues, and their comments will undoubtedly have an impact on the high-level decision-making.

Since 1994, Jiang Ping has gone abroad annually to give lectures. In 1983 when he made a study tour as a scholar, a foreign professor who received him sighed that "China has no legist." Now, he is a guest welcomed everywhere, with an honorary JSD degree and the title of honorary law professor of the Ghent University of Belgium, Catholic University of Peru and the Tor Vergata of Italy. He is the third Chinese with an honorary JSD degree granted by prestigious foreign universities, after Song Chingling and Deng Xiaoping, and the first scholar rather than a statesman to be given such an honor. Jiang Ping regards the honor as acknowledgement by foreign countries of China's promotion of the rule of law over the decades.

Now, Jiang Ping has difficulty walking. Every morning, he has a walk with his dog, and then reads books. In the afternoon, he often receives guests. He receives any student of China University of Political Science and Law. It's a long distance from China University of Political Science and Law to Jiang Ping's home, but there is a non-stop bus. Students all know that at the terminal is their teacher.

"Old Shanghainese" Gets Lost in Pudong

Zhang Yiqing: | 1978: Worker in the Sixth Shanghai Textile Machinery Plant
2008: Retired from the Yanxing Group

Standing in Lujiazui in Shanghai's Pudong, once so familiar to her, Zhang Yiqing felt lost. It had changed so much, with so many high-rise office buildings of foreign-funded companies, brand name hotels and shopping malls.

Thirty years ago, she went across the Huangpu River between Pudong, essentially a farming area east of the river, and Puxi, the urban area of Shanghai west of the river, several times a month. Because her father had difficulty in moving about, she had to fetch her father's wages and deal with trivial matters for him. At that time, the only way to get across the Huangpu River was via ferry. People all wore grey or blue clothes, even the styles were similar. One day she met a friend on the ferry. They felt close because they were wearing the same clothes. But next time, when they met again on the same ferry, they were still wearing the same clothes. This time they viewed each other with an embarrassed smile.

During a winter in the 1970s, Zhang Yiqing went as usual to the ferry dock at Yan'an East Road in Puxi. The ferry fee was six fen (cents). Everybody's wages were similar then, about 36 yuan. This six fen was the round-trip price from Pudong to Puxi. Zhang Yiqing always boarded the ferry at Yan'an East Road and landed at Lujiazui in Pudong. Along Pudong Avenue there were only two bus routes — Route 81 and Route 85. This was the only road in Lujiazui at the time. Zhang usually took the Route 81 bus and got off at Liuhaoqiao Stop.

That day, however, was unusual. Since she had been delayed at her father's work unit, it was already 10 p.m. when she left it. Today's Pudong is full of shopping malls, their lights making the night as bright as day. But back then, there was only darkness, even the street lamps were dim. There were no shops at all and only a few passersby. Zhang Yiqing was afraid. Unknown creatures sounded once in a while in the fields, which extended one after another from the road for some distance. Even worse was that it was too late for the Route 81 bus.

Zhang Yiqing had to get home. In her father's old brown coat, she walked quickly, not daring to turn her head. At first, there were several female workers around who had come off the night shift. Zhang Yiqing followed them. But before long, these women disappeared. There was only Zhang Yiqing on the road. She was even more afraid, and automatically quickened her steps .

She finally arrived at the dock at Lujiazui and boarded the ferry. Ten minutes later, she had returned to Puxi. Though Puxi was much busier than Pudong, it still couldn't compare with today's bustling Bund. All buses had stopped, and there were few passersby. Only the street lamps seemed brighter than those in Pudong. Having no alternative, Zhang Yiqing continued walking. It was 2 a.m. when she arrived home.

Her mother was still standing at the door, anxiously looking down the street into the distance. She finally set her heart at rest when she saw her daughter returning. The mother had thought of calling the father's work unit. At that time, there were no private telephones in citizens' homes, and the mother would have to go to the phone booth at the entrance to the lane. It was too late that day; the person manning the booth would have already left work.

Later, Pudong experienced a great change.

Around 1990, a development program for Pudong was initiated. The plan was to set up the Pudong Development Zone, Pudong Industrial Zone and Pudong Economic Town. The improvements in transportation and traffic conditions reflected the progress being made in Pudong. In the early 1990s, the Nanpu Bridge, the first double-tower cable-stayed bridge designed independently by the Chinese, was erected, and opened to traffic on December 1, 1991. This bridge, with a three-level spiral interchange approach in Puxi reaching to the Inner Ring Overhead Road, the South Zhongshan Road and the Lujiabang Road, connected with Pudong at the other end. Later, the Yangpu Bridge, another double-tower cable-stayed bridge across the Huangpu River, was erected. In the late 1990s, the project of the northern and southern tunnels across the Huangpu River was completed, creating a main route for people to cross the Huangpu River.

With Pudong developing at high speed, the ferry was no longer the only means of getting across the Huangpu River. Since her father had retired, Zhang Yiqing no longer needed to go to Pudong on a regular basis. In November 1997, Zhang Yiqing's family, who had lived on a lane in the Shikumen (stone gate houses) style featuring a courtyard for over four decades, moved to the northeastern part of the Hongkou area.

In fact, the move of the Zhang family was prompted by the commercial construction in the Hongkou District.

North Sichuan Road is the third largest commercial street in Shanghai next to Nanjing Road and Huaihai Road. This road, stretching 3.7 km across the Hongkou District, and running from north to south, was one of the earliest roads built when Shanghai opened to the outside world as a port. In the 1990s, it was redeveloped as a "popular commercial street." In accordance with the plan, the newly-built North Sichuan Road provides customers with high-quality, middle- and low-priced products and electronic services. It features cultural events and recreational establishments in the adjoining areas. The target customers are salaried workers. It is different from Nanjing Road which features many famous brands and Huaihai Road which features luxurious goods. On this road, products are mainly from domestic enterprises. A billboard was once erected on the road, saying: "Other roads may be good for people looking around; if you want to buy something, Sichuan Road is your best choice." Many commercial buildings, including the East Building of the Shanghai Seventh Department Store, Duolun, Kailun, Shanghai Spring, have beocme the landmarks of this commercial street.

Zhang Yiqing's old house was on the North Sichuan Road. The ground floor of the building facing the street had many shops, such as the Daxiang Food Shop, the Seventh Department Store and the Yingzi Photo Studio. So, naturally they and other residents on the commercial street were asked to move elsewhere.

For Zhang Yiqing, this move was wonderful. Their former home in North Sichuan Road was opposite to noisy Shikumen houses. It had no elevator, which made it inconvenient for Zhang Yiqing, who had developed rheumatoid arthritis in middle age, to go up and down the stairs. It was not equipped with flush toilet. Safe use of electricity and dependable supply of water were also a problem. In the 1980s, they needed to keep the tap turned on during the night to collect water for the next day's use. The new apartment has all the modern amenities.

Because she had developed lumbar disc disease, Zhang Yiqing enjoyed an early retirement in 1998. When her daughter went to study at the university in 1999, she felt a sense of void. Since she had a mind eager for lifelong learning, she enrolled in a computer course offered in her community. She knew nothing about how to enter Chinese characters, and in the beginning she couldn't even remember the 26 letters on the keyboard, but still she kept on studying. She felt it was interesting to gain new knowledge in a totally new field. After a period of hard learning, she passed the national elementary exam on computers. In her spare time, she surfs the Internet on the old computer her daughter once used.

Besides learning to use the computer, Zhang Yiqing loves going to theatres. She has always gone to the Shanghai Majestic Theatre, the Shanghai Art Museum and the Shanghai Museum to watch various performances and exhibitions. Ballet is her favorite. Whenever the Shanghai Ballet or overseas ballet troupes have performances, she goes to watch whenever her health permits.

The establishment of a series of new theaters like the Shanghai Grand Theatre further satisfies Zhang Yiqing's need for art performances. Whenever Zhang watches plays in the Shanghai Grand Theatre, she can't help remembering the scenes from 30 years ago when she watched the ballet performances *Dagger Society* and *Swan Lake* on a simple wooden chair in the city council auditorium. Although it has been years since the auditorium was pulled down, she is still touched by her beautiful memories.

Since her daughter has graduated from university, Zhang Yiqing sometimes goes to watch performances in the company of her daughter. But as her daughter and son-in-law are quite busy with their work, she usually goes with friends.

Not long ago, Zhang Yiqing went to the Shanghai Oriental Art Center in Pudong to watch a performance. This was the first time in many years that she had come to the district of Pudong. Though she had once been so familiar with this district, she now felt lost in this totally changed area. Facing the Lujiazui Financial Zone, she imagined seeing crop fields superimposed on the high towers. Seeing the Oriental Pearl TV Tower, the Chia Tai Mall and the Pudong Shangri-La Hotel, she smiled in great excitement and happiness...

"Crazy" about Lepers for the Rest of Her Life

Li Huanying	1978: Chief physician of the Dermatopathy Research Institute under China Academy of Medical Sciences; transferred from to the Tropical Medicine Research Institute under Beijing Friendship Hospital
	2008: Researcher of Beijing Tropical Medicine Research Institute, councilwoman of International Leprosy Association, vice-chairperson of the China Leprosy Association

In 1970, Li Huanying came into contact with lepers for the first time at an isolated leprosy village in Taizhou, Jiangsu Province.

Leprosy is a disfiguring, contagious, chronic disease. It causes ulcers at many places on the patient's body and may result in disability. During the period of the Republic of China (1912-1949), there were no effective measures to control this disease due to backward medicine, and the government adopted the policy of isolation, sending lepers to remote villages for treatment. After New China was founded in 1949, leprosy villages were retained as special organizations for isolated treatment and centralized management.

In the half year during which she stayed in the leprosy village, Li Huanying came to understand the fear and discrimination hindering the treatment of leprosy. In 1978, she selected the difficult topic of leprosy prevention, although she was about to retire. She expressed her wish to the head of Beijing Friendship Hospital: to once again select a research topic and specialize in leprosy prevention. At the age of 57, she was transferred to the Tropical Medicine Research Institute of Beijing Friendship Hospital and started her second career. Yunnan, Guizhou and Sichuan had a high incidence of leprosy, where the number of lepers accounted for about 70 percent of the national total. In the autumn of 1979, Li Huanying went to Yunnan alone. Nanxing Village in Mengla County, Yunnan Province was a village with the most serious epidemic of leprosy among all the places she had visited. To reach her destination, she first spent three to four days on a train from Beijing to

Kunming, capital of Yunnan Province, then took a long-distance bus to the county center and finally walked six to seven hours along mountain paths, which were narrow trails mostly walked on by oxen and horses and allowing only one person to pass. On some sections, a mountain stood on one side and a cliff on the other. For fear of contracting leprosy, no doctor was willing to work there voluntarily, and most patients were not able to come out after entering the village. All things used by lepers needed to be disinfected. A doctor would wear insulated clothing, a pair of rubber shoes, a pair of gloves and a gauze mask, stand a meter away, hold a stick with pills at the other end, and then guide patients to take the pills with directions expressed in exposed eyes. Without specific drugs, lepers suffered various disabilities, such as broken hands or legs, flat noses, blindness , and a ferocious appearance, which frightened others. All these evoked sympathy in Li Huanying for the lepers. An eight-year-old boy led his blind mother by the hand and came to Li Huanying. The mother held out her hand and withdrew it immediately three times. Li Huanying took her hand and said, "I'm a doctor and have come to treat you. I'm not afraid." She knew that leprosy could be prevented and treated and was not frightening, and that doctors did not need to wear insulation clothes when treating lepers. Therefore, she shook hands with lepers, hugged them, patted them on the shoulders and even stroked them without wearing insulated clothing. In this way, she was conveying a message to the village cadres: "I don't fear them as a doctor. What are you afraid of?"

In 1980, she went to the United States among the first group of visiting scholars from China after the reform and opening-up policies were adopted. The visit lasted a year. It was the first time that she was united with her relatives in the United States after she had returned to China. Her parents had passed away, and her younger brother and sister tried to persuade her to stay in the US: "Sister, you're growing old and are alone in China. Stay here with us." However, she vividly remembered her promise to the lepers: "I will bring back medicine and cure your disease." She investigated seven international leprosy centers in the United States and the United Kingdom and learned about the advances in international leprosy prevention, aiming to change the situation in China whereby lepers were isolated for a long time. Before she returned to China, she applied to the World Health Organization (WHO) for new anti-leprosy medicine for 100 patients. Before the Spring Festival in 1983, Li Huanying started to set up a pilot short-term outpatient clinic combined with chemotherapy at Nanxing in Mengla County, Yunnan Province, the first of its kind in the world. She had close contact with lepers and moreover encouraged local officials to do so. Once, she shook hands with a leper and then guided that disabled patient to a local official and said, "Come on, you two shake hands." The official managed to touch the patient's hand though he was reluctant. In this way, with a smile Li Huanying taught a lesson to all the people: Leprosy is not terrible. It can be cured with timely therapy.

A tent was set up to serve as an operating room; a worktable was made of bamboo. Under basic and hard conditions, Li Huanying devoted herself to her work and tried to gain experience. The short-term treatment combined with chemotherapy recommended by the United Nations requires that the maximum period for taking medicine should be 24 months. With no successful experience to refer to, she was unsure of what would happen. After patients took the medicine, they began to have serious pigmentation changes on their faces, looking horrific in black and purple. She was afraid that patients would stop their treatment. But they trusted her, and persisted in taking the medicine even if their faces turned black. Finally 24 months later, the experiment succeeded. Those recovering from leprosy told her: "Leprosy turned us into ghosts and you turned us back into human beings."

Two years later, the 47 lepers who had taken the medicine all recovered. Later, the same experiment also succeeded in other provinces. Her success shortened the period of leprosy treatment by three and a half years. Based on this, she applied to the WHO for funds to expand the pilot to 59 counties around China. Several decades later, she had cured more than 10,000 lepers, with a recurrence rate of only 0.03 percent, far lower than the 1 percent allowed by the WHO.

In her office, there is a transportation map of Yunnan, Guizhou and Sichuan on which there are over 20 small red flags marking the places where she has been. If she went to a place twice or three times in a year, then she would have been to these places at least 40 or 50 times in over two decades. However, she understands that she cannot eradicate leprosy by herself. Since 1985, she has been giving lectures at classes to personnel engaged in leprosy prevention in Yunnan, Guizhou and Sichuan. Each class has 40 people. Altogether over 10,000 people have attended her training classes.

In 1989, Li Huanying went alone to the leprosy village in Xichang, Sichuan Province to collect samples. On her way back, an accident happened when the car climbed over a snow-capped mountain. She was immediately sent to hospital. Her left clavicle and three ribs were broken, and her head was injured too, requiring seven stitches. Since she was single and her relatives were abroad, her assistant wanted to stay with her and take care of her. However, Li Huanying asked her assistant to immediately return to the laboratory with the over 200 blood samples and to finish the experiment. While she was in the hospital, she worried about her experiment and thus often slipped to her office and typed an English report with the hand that could move. When engaged in her work, she was concerned about nothing else. She has been in danger four times, with many scars on her body. When her bus overturned, she said, "It should be my turn anyway if counted on the probability of my taking the bus." When her boat turned upside down, she said, "I'm as fat as an inflated ball, and will not sink." This time, although she suffered fractures in the left clavicle and three ribs, she still continued to work hard.

April 17, 1990 was a most memorable day for Li Huanying, and the happiest day for residents of the Nanxing Village, Mengla County, Dai Autonomous Prefecture of Xishuangbanna, Yunnan Province. On that day, the county head declared that Nanxing would be renamed Mannanxing and be included into Mengla County as an administrative village. Mannanxing means "a new riverside village" in the Dai language, signifying that this once isolated village was once again reunited with the human world. Young and old people in the village made Li Huanying put on Dai costumes and wear flowers, and Li was so excited that tears came to her eyes. Her seven-year efforts had finally been rewarded. Today, the average income per household has increased to several thousand yuan; a road extends to the entrance of the village; a satellite television ground station has been built in the village and all households have TV sets; all the thatched roofs have been turned to tile ones; some families have even got walking tractors.

In 1998, the Fifth International Leprosy Conference was held in Beijing, at which the special action plan consisting of short-term therapy combined with chemotherapy headed by Li Huanying and aiming to get rid of leprosy was acclaimed as "the best treatment action in the world" and she was elected executive chairperson of this conference. In 2001, the research on the strategies, prevention technologies and measures controlling and basically eliminating leprosy in China headed by Li Huanying won the State Award for Scientific and Technological Advances, first class. In 2002 when Li Huanying was in her eighties, she launched a research project on the molecular epidemiology of leprosy with the goal of basically and finally completely eliminating leprosy, which was called "the last battle" by the international academic circles of leprosy. In November 2005, Li Huanying Medical Foundation was established at Beijing Friendship Hospital, and she would donate her savings and domestic and foreign sponsorships to encourage young people who have made outstanding contributions to the scientific and academic research of leprosy.

Her advocacy has fundamentally changed the situation in which lepers and leprosy villages were marginalized. Starting from the 1980s, a lot of leprosy villages around China were closed and reorganized, while over 500,000 cured lepers walked out of the leprosy villages, found employment or enjoyed a peaceful life in their later years. Leprosy villages are no longer synonymous with impoverished villages. At each village, there is a patients' autonomous organization and an administrative committee whose members are elected by villagers through democratic election, which take charge of village production, financial management and means of production, patients' lives and cultural activities, and assist professionals in medical and rehabilitation work. Lepers can become engaged in farming and sideline production as ordinary farmers do, and can thus earn their living and are basically self-sufficient. Moreover, governments at various levels give a lot of material subsidies to these villages; consequently, lepers are adequately fed and clothed, and enjoy a rich cultural life with such facilities as schools, parks, clubs, gyms and libraries. Some leprosy villages were renamed "happiness villages" and some cured patients voluntarily stayed there.

In 2007, on International Leprosy Day, the Ministry of Public Health of China announced to the world that the number of lepers in China had decreased from about 520,000 in the late 1940s to around 6,300. The figures marked the fact that China had left leprosy behind after over 50 years.

A Globetrotting Chinese Journalist

Kong Xiaoning:
1978: A student of the School of Journalism of Fudan University
2008: Director of the Travel Department of the *People's Daily Overseas Edition*, senior editor and member of the China Writers' Association

Kong Xiaoning likes travelling. Travel can refresh a person's mind and allow one to experience different cultures. He is also a man with literary talent. Through his descriptions, dull words become lively and vigorous. He is the first Chinese journalist and one of the very few in the world who has ventured to the North Pole, Antarctica and the Qinghai-Tibet Plateau (the so-called "third pole of the earth") for the purpose of reporting news.

In the spring of 1995, as a *People's Daily* journalist, Kong was chosen to join the first Chinese North Pole scientific expedition and report on the entire journey. Overjoyed as he was, he was more concerned about whether he would be able to keep up with the research team and do a good job in covering the stories. Before his departure, he participated in the strict simulation training on the frozen Songhua River at minus 40 degrees. The training was intensive and hard, but Kong never complained or thought about giving up, because this opportunity was so rare and important. Not many journalists were lucky enough to be given such a chance.

Finally the day came. At 3:20 pm Beijing Time on May 6, 1995, a plane landed on the North Pole. As soon as the extension stairs were ready, no one could wait any longer and everyone rushed out of the plane. Standing on the ice of the North Pole, they stretched out their arms and hugged each other excitedly. Kong and the other 15 members unfolded a huge national flag, facing the Tian'anmen Square, which is more than 10,000 km away from where they were standing.

This flag was handed to them by the Tian'anmen Square Guards of Honor before they departed.

The following year, Kong went to Antarctica. During his two months in the Zhongshan Station (a Chinese research base on Antarctica, first built in 1989), he worked as a journalist and a part-time laborer. Kong was even responsible for cleaning and painting the outer face of the station. After he returned home, he wrote a book titled *Adventures in the Wilderness of the Antarctica* about his experience with the 13th Antarctica Scientific Expedition. This book won the most prestigious national book award of the year — the "Five-One Project" (an annual award organized by the Chinese Publicity Department; and "Five Ones" refers to one best drama, one best TV show, one best movie, one best book and one best academic paper recommended by each province to enter national selection).

Before this, Kong also followed a scientific expedition to the Hoh Xil uninhabited zone on the Qinghai-Tibet Plateau, one of the wildest places with breath-taking natural beauty on earth.

He did not become a war correspondent or a political correspondent, but his rare experience of having visited the three most-difficult-to-reach places in the world has realized a dream about becoming a journalist which Kong had as a youngster.

In 1980, after Kong graduated with honors from the School of Journalism of Fudan University, he got a job at the Wuhan University in Hubei Province to teach journalism course. However, for some reason, the plan to establish this course was dropped. Kong was then transferred to the school's publicity department to be the editor of the school newspaper. Half a year later, he was enrolled into the Department of Journalism in the Graduate School of the Chinese Academy of Social Sciences. His tutor was the chief editor of the *Market Daily* under the *People's Daily*. His tutor was very appreciative of Kong's talent and decided to invite Kong to work with him.

His "big newspaper" complex has contributed to Kong's decision to join the *People's Daily*, which has always been the foremost authoritative and influential newspaper in China. From the end of the 1970s to the early 1980s, there was a wave of people in China who went abroad. Meanwhile, the relationship of the mainland with Taiwan and Hong Kong showed delicate changes and China's work of overseas Chinese affairs showed new characteristics. The original edition of the *People's Daily*, which was mailed from China, could no longer satisfy the needs of overseas Chinese readers. Therefore, the overseas edition was born. It was edited in China and printed and distributed directly in overseas cities with a large Chinese community. On July 1, 1985, when the first issue of the *People's Daily Overseas Edition* came out with Deng Xiaoping's inscription: "Greetings to Overseas Chinese Friends," Kong had developed into a qualified editor and reporter for the newspaper's Education & Science section. Since then, he has been writing travelogues.

The *People's Daily Overseas Edition* is the first domestic newspaper to use computer-aided typesetting. This first computer system used by Chinese newspapers was introduced from Japan. It was slow and inconvenient, with several thousands of commonly used Chinese characters scattered onto different plastic discs. The typist often had to look all over them for the right characters. After typing, layout and getting the printing paper ready, workers had to make films. The film was then cut-and-pasted into full page galleys used to create plates for offset printing. If they found mistakes, they had to scratch the wrong characters off with a little knife and stick the right ones onto the paper. It was all done by hand, which took a long time. At that time, the newspaper staff often worked until two in the morning. Then they ate midnight snacks in their dining hall, and went back to work until three or four am. Even though the conditions were unfavorable, Kong and his colleagues were very content, because it was already much easier than the original procedure.

For a month, Kong had worked as an intern in an old typesetting workshop. The workshop was packed with wooden boxes piled as high as an average male adult. These cast lead sorts were placed in type drawers in the order of their radicals. The compositors had to remember exactly where each Chinese character was kept. It was very common to see a compositor holding a manuscript in one hand while picking the right characters out of different drawers with the other hand. In summer, the workshop was always filled with the smell of lead and with hot suffocating air. It was almost intolerable. When there was urgent news, workers would be specially arranged to deliver the journalist's report to the typesetting workshop as the journalist finished each page. At those times, more than one compositor would look for Chinese sorts at the same time. However, even with many people working hard at the same time, the work efficiency didn't increase that much. By the time the newspaper was finally printed, it would already be daybreak.

At the end of 1980s, the newspaper started to use the Founder System created by a Chinese named Wang Xuan, with which all editing work was done on the computer, greatly reducing the workload.

As the typesetting system continues to be upgraded, the efficiency of editing and publishing has been greatly increased. Advanced technologies facilitate news gathering and transmission. In the past, Kong only had a notebook and a pen with him for news gathering. He had to borrow a camera from his office if needed. If he was out of Beijing, articles could only be sent back to the office via telegraph. Fax machines didn't appear until the 1990s. But now Kong carries his laptop, recorder and digital camera to work. Some journalists even have a digital module, satellite telephone, TV relaying vehicle and mobile workstation.

In the meantime, the media circle and content were also undergoing drastic changes. When Kong was in college, there were only a handful of universities that offered courses of journalism. This major attracted all the top liberal arts students in China. The *People's Daily* and other state-owned media had a concentration of the very best students and even for a time served as training centers for preparing government officials. During Kong's first few years as a journalist, the *People's Daily* was centered on the Party and government affairs such as Party work experience, achievements and typical examples for other parts of the country to follow. It played a very important role in the Party organizations. Therefore, when an ordinary journalist from the *People's Daily* went to do reporting in other cities, he would be warmly greeted by high-ranking local officials and given special treatment. When the *People's Daily Overseas Edition* was launched, it quickly drew a large number of overseas readers with its new content which reflected modern China's development and ordinary Chinese people's daily lives. It received wide recognition from not only the higher authorities but also overseas readers, as well as other media. A lot of its published news reports and columns have won national news awards. Many newly established city newspapers have followed the style of the *People's Daily Overseas Edition*.

In recent years, new magazines, newspapers and other new forms of media have sprouted to meet readers' diversified needs. Interviewees sometimes prefer a more visualized medium like TV or a quicker and more direct medium like the Internet. The state-owned *People's Daily* no longer holds a superior position in the media world. It faces fierce market competition. The *People's Daily Overseas Edition*, which used to be the only source about China's mainland for overseas readers, is also losing its dominant role in the overseas market. Newly-established overseas Chinese newspapers, TV and websites are taking more and more of the market share. As a result, the people working in the media industry feel that they are shouldering more responsibilities and are under more pressure than ever. However, all of these challenges have never upsetted Kong. Kong never complains and actually quite enjoys his work. He believes that his job as a journalist has a much more important goal than to earn money; that is, to record social changes and people's lives. This requires a journalist to be keen to interesting stories and write everything he knows from a unique angle.

He has published more than ten books with over two million words in total. His column has been popular among readers, and many overseas Chinese newspapers reprint his articles without any abridgement.

So far, Kong has traveled to all continents except South America. He has written a book called *Footprints All Over the World* which includes all of his articles from the past ten years. At present, he continues to work hard and exercise every day. He believes that he won't have to wait much longer to reach his last destination, South America.

From a "Migrant Worker" to an NPC Deputy

Kang Houming

1978: Farmer from Shiniu Village of Laisu Town in Chongqing, Sichuan Province

2008: Leader of the migrant workers team of the No.1 Municipal Engineering Company of the Chongqing City Construction Holding (Group) Co. Ltd., and deputy to the National People's Congress

In the spring of 2008, the 45-year-old Kang Houming, a migrant worker from Chongqing, attended a session of the National People's Congress (NPC) held in the Great Hall of the People in Beijing as a deputy. This was the second time he had been to the Great Hall of the People. The previous occasion was in 2005 when he was given the award of National Model Worker. Three years later, he stepped into the Great Hall of the People again, this time for the 11th NPC. Among the 2,987 NPC deputies, Zhu Xueqin from Shanghai and Hu Xiaoyan from Guangzhou were both migrant workers, just like Kang Houming. This was the first time that migrant workers had been elected as NPC deputies. Technically, they represent their constituencies, but almost everybody regards them also as representatives of a social group loosely termed "migrant workers," although such a term cannot be found among the classification of professions and trades. It is estimated that they exceed 200 million in China.

The term "migrant workers" refer to those with registered permanent rural residency who work for companies in urban areas. Because of the residency system in China, they cannot benefit from the same medical and employment insurance as urban residents; nor can they enjoy the social benefits which is a result of urban economic development. And there was no migrant workers' union that could voice their opinions on their behalf. Migrant workers constitute the largest single group among industrial workers. Compared with other groups, they have an unfavorable work environment, heavy workload, and low pay, and are not particularly respected by other people. Not until recent years has the Party Central Committee and the State Council started to pay attention to this group and to specifically use the term "migrant workers" in government documents addressing the need to solve their problems. Since then this term has been widely used.

As great importance has also been attached to "migrant workers" in recent years, Kang Houming and the other two migrant-worker NPC deputies became the media's most popular interviewees. Kang Houming says that it is more tiring to give interviews than to work. Their primary duty as NPC deputies is of course to participate in the deliberations of state affairs. On February 18, Kang Houming received a formal invitation to attend the NPC session. At that moment, there were more than 20 workmates on the scene. All of them felt very happy for him. They all wanted to have a look at this invitation. He was going to Beijing on February 28, but before that he still had to continue working at the construction site and as his son was ill, he could only spare two days to gather his colleagues' opinions and requests. As Kang Houming was a migrant worker just like everyone else rather than an official, his colleagues all spoke freely and Kang carefully wrote down everything. He had never felt so proud because he was about to report to the Premier. At the NPC meeting, a vice-premier met with him and carefully listened to his account of the migrant workers' opinions and problems. The vice-premier asked him, "Is everything you have said true?" Kang Houming answered, "Yes." Then the vice-premier said, "Good, I've recorded everything. Thank you very much!"

After the NPC session was over, Kang returned to Chongqing and continued his work at the construction site as before, toiling to support his family — he has been doing this for 28 years.

Kang was born into a farmer's family in Yongchuan. His family had not been given enough farmland, so he was not needed at home. After he graduated from middle school in 1980, he left his hometown to make a living on his own. His first stop was Guangzhou, but he couldn't find any job there. He didn't know how to get a temporary residence card, so he had to hide at night to avoid police checks. Later on, he went to Tibet and Chongqing. He tried many different jobs, from food processing, stonework to construction work. At the beginning, he only earned 1.8 yuan per day and the worst thing at that time was that quite often construction companies delayed payment. On one occasion, Kang's boss owed the workers over 10,000 yuan. When he went to demand their pay together with his colleagues, they discovered that their boss had already run away. Later they found out that the company had gone bankrupt . Kang learnt his lesson from this experience and from that time on he determined to never work for an unqualified company again. And whenever he heard that anyone had chosen a company based only on pay, he would warn them that they should find a credible and qualified company to work for. Only in this way would their living conditions, remuneration and accommodation be ensured. The most important thing is to choose an honest, credible employer. After all, there are now more and more of them.

In 1998, Kang became a construction worker for the No.1 Municipal Engineering Company of the Chongqing City Construction Holding (Group) Co. Ltd. He has been doing construction work such as paving roads and building bridges for about ten years. "Wherever the construction site happened to be, that was our home." Kang still remembers that when he just started to work for this company, he lived in a shabby hut. Later he moved into a somewhat nicer shed which accommodated thirty to forty people. Not until 2003 did his life of migration finally end. His company built a migrant workers' residential building at Maoxiangou, with each room being six meters long and four meters wide. Each room had six bunk beds and was home to 12 workers. Two years later, their rooms were installed with air-conditioners and 24-hour hot water.

The team which Kang heads has ten married couples, but the company can only provide collective housing for them. If a couple wants to live together, they have to go out and rent a place. However, the cheapest room is 200 yuan per month. These couples came up with an idea. They separated the shared room into several compartments with thin wooden sheets. These "homes" were very inconvenient for married couples. However, the good news is that the Chinese government has already built a lot of cheap rental houses all over China. Cheap rental housing is the Chinese government's housing policy intended to subsidize and provide cheap rental apartments to those who live on the minimum wage and can't afford to rent a flat. Although Kang's dormitory has fairly comfortable conditions, he still hopes to move into one of these cheap rental apartments. Every time he finishes paving a new road, he looks up at the high-rises and dreams of buying his own apartment. Of course, there is still a long way to go before Kang can fulfill this dream.

Recently, Kang got another pay raise. Now he earns 2,000 yuan per month. He receives a raise almost every year. His wife is a cook at Kang's construction site and makes 800 yuan every month. For a family which has to support a college student, a monthly income of 2,800 yuan is insufficient. Fortunately, two years ago the government regulated that every enterprise must provide their employees, including migrant workers, with medical and work-related injury insurance. Since then, Kang no longer needs to worry about hospital expenses. His child majors in computer programming, a coveted subject in China. Kang has put all of his hopes on his child. He believes that everything will become better when his child graduates from college and starts to make money.

Although Kang is still a migrant worker, his five-year tenure as an NPC deputy has just started. In front of the media, Kang has said a lot, but he never forgets to say "I'm here to learn." During his spare time, he conducts surveys among his colleagues and sometimes communicates with company leaders and governmental officials. One month after he returned home from Beijing, he was "apprenticed" to a professor of the law school in the Southwest University of Political Science and Law in order to improve his ability to participate in deliberation of state affairs and thus to better serve migrant workers. After the professor accepted Kang as a "student," Kang said that he now had more confidence in his role as an NPC deputy.

His urban life as a migrant worker continues. About two thirds of the young people in Kang's hometown have left their hometown and work in cities, and most of them are married couples. Young countryside people hope to start their careers in big cities. Kang wouldn't mind going back to the countryside once he is old. In fact, whether a migrant worker goes back to his hometown or not is generally determined by how much money he makes. If he has his own career and makes a lot of money in the city, he will likely choose to stay in the city. His second best is to buy, if he has enough money, an apartment in the county seat; or he has to do with one in his native town. Anyway, his life has become much better than before and will hopefully continue to improve in the future.

Introducing Tibetan Culture to the Interior

Sonam Tenzin Qopei

1978: Chengguan primary school pupil in Lhasa, Tibet
2008: President of the Tanggula Cultural Dissemination Co., Ltd. and vice-chairman of the Tibetan Young Entrepreneurs' Association

"Spread Mount Tanggula's sacred name and Tibetan culture to the interior of our country." Whenever Tenzin Qopei speaks of this, he sounds very determined and proud. The interior part of China is his second home and has brought him a rich and fabulous life experience.

In 1978, the seven-year-old Tenzin Qopei was enrolled into the Chengguan Primary School in Lhasa and started his education by studying Tibetan letters. Although Tenzin was in Lhasa, the capital city of the Tibet Autonomous Region, at that time, Tenzin had never seen things like colorful pencils, pencil boxes, magic cubes and cassette recorders, which are very common in Tibet at the present time. For Tenzin, that period of his life was purely about acquiring knowledge. He was very eager to learn and did very well in every subject in school. In 1986, he was sent to the Sixth High School of Shashi in Hubei Province for junior high school education along with several other top students. It was his first visit to the interior and also true of most of the 51 other Tibetan boys and girls. Living among Han students, he and other Tibetan students did not feel uncomfortable at all. There were also a few Tibetan teachers accompanying them to the school to take care of their daily needs. The school not only provided everything for them, including free books, stationery, accommodation and food, but also gave them money on a monthly basis. During the day, they studied and exercised with Han students; at night, teachers would help them with homework or organize activities for them. On weekends, the school organized outings for them, such as going shopping, visiting libraries, museums and cinemas.

After Tenzin graduated from high school, he was recruited to the navy. On his first day in the navy, his instructor paid special attention to this Tibetan newcomer. He told Tenzin to let him know if he had any problems or needed anything. Tenzin remembers that during his high school years the teachers gave him extra attention because he was from a minority group. He thought that he might be adding trouble to the school. Therefore, when the instructor tried to give him special treatment, he said, "I've been living in the interior for several years. I'm already used to the lifestyle here. So please don't give me any special treatment just because I am a Tibetan. To protect our country is not only the duty of the Han people. It's also the duty of all Tibetans. I thank you, but I really don't need special care." Time passed very quickly, and soon Tenzin became a qualified naval seaman and had made friends with everyone. His military unit often held parties on festival days and his Tibetan performance was always a must-see part of the programme. On every Tibetan New Year's Day and on other Tibetan festivals, his superiors would see to it that his comrades would join him in celebrations. Special food was also prepared for him, which he usually shared with his comrades.

In his second year in the navy, a few vacancies opened up in a military university. Studying at a military university is every soldier's dream. When a superior officer came to identify the most qualified candidates, he was surprised to see that everyone recommended Tenzin. In that year, Tenzin entered the Chinese Institute of Naval Engineering of the People's Liberation Army and started his college education. After he was later discharged from the military service, he was assigned to a government office in Lhasa. Everyone including himself thought that a successful official career was just beginning. But his life took an unexpected turn.

Today Tenzin is no longer a government official. He runs his own Tibetan trinkets business. He wears a heaven's-bead bracelet. This bracelet is very delicate and every bead is very dark and polished so that it is very shiny. The middle one is especially pure and bright, looking almost holy. On one occasion, a real estate developer was choosing heaven's beads at his store, but then he saw the bracelet on Tenzin's wrist. He liked it so much that he wanted to buy it from Tenzin. At first, he offered 10,000 yuan, but Tenzin refused to sell him the bracelet. Then he offered up to 100,000 yuan, but Tenzin refused again and said, "I would never sell this bracelet no matter how much you might offer me for it." Tenzin has had this bracelet for years and has never taken it off. It serves as a testament to many trips he has taken between the interior and Tibet.

In 2002, Tenzin turned 31. He started to think more seriously about his goals. He is an adventurous and innovative person. He believes that one should do something different from others. What would be his dream? He thought of his second home — the interior part of China, where he has a lot of good memories. To blend both cultures of Tibet and the interior into every moment of his own life has always been Tenzin's wish. Upon reflecting on this, Tenzin chose a clearer goal for his life: to spread profound Tibetan culture to the interior. However, this goal seemed too big and vague. "But where shall I start?" Tenzin unfolded a map of China and found that the city of Jingzhou (in 1994, the two districts, Shashi and Jingzhou, were amalgamated into the city of Jingsha, and later it was renamed the city of Jingzhou), where his high school is located, is in the centre of China geographically. Therefore, he decided to choose Jingzhou as his first destination.

Several days later, Tenzin collected some examples of Tibetan-style furniture, a few jars of barley wine and butter tea, some Mani stones and Tibetan craftwork made by Tibetan people. He took all of them with him from Lhasa to Chengdu by air, then transferred from Chengdu to Wuhan. Afterwards, he took a bus to Jingzhou from Wuhan. Tenzin remembers very clearly that back then the air freight charge was 27 yuan per kilo. Because there was no direct flight, it took him six days to get to Jingzhou. He opened a small teahouse on a street close to his high school to teach customers about Tibetan culture while selling tea. At that time, the Tibetan class in Jingzhou Sixth High School was still open, but there were many more Tibetan students than before. As Tenzin shuttled back and forth between Jingzhou and Lhasa, every time he left Lhasa, Tibetan parents would always ask him to bring things to their children in Jingzhou. Amulets were always an indispensable item. These simple but meaningful handmade amulets carry endless family love, and draw a lot of interest from customers who see them in his teahouse. When more and more Chinese asked him to buy amulets for them, an idea struck Tenzin: why not sell Tibetan ornaments here?

Soon he opened a Tibetan knick-knack store named "Tanggula" and it quickly became very popular in Jingzhou and throughout Hubei Province. The capital city Wuhan was planning to build an ancient-style commercial street. The leaders of the Provincial Department of Nationality Affairs encouraged Tenzin to open his store on this street. When the commercial street organizing committee learned that "Tanggula" planned to open a branch store on the street, they gave Tenzin preferential treatment — free rent and no taxes for three years. He was very happy to hear that news and quickly started his business in Wuhan.

In Tanggula, none of the products have price tags. Only when a customer has decided to buy a product will Tenzin indicate its price. People are often attracted to the store by the unique Tibetan folk songs played in the store. Tenzin never brags about how good his products are. Instead, he is always patient about describing the delicacy of Tibetan craftwork and the profound nature of Tibetan culture. There was an old man who visited his store regularly, but he never bought anything or asked about the price of any item. He just looked around and listened. Tenzin asked the old man why. The old man told Tenzin that he was a Wuhan University professor. He said, "All the things you explain to the customers would take me about two months to research and understand. But it only takes me 20 minutes to understand all of them by listening to you."

Tenzin has now opened a lot of Tanggula branch stores. People from Beijing, Shanghai and Guangzhou asked to join his business chain. Soon the original business mode of "collecting products from Tibet and selling them in the interior" could no longer meet the customers' needs. In 2005, Tenzin established the Tanggula Cultural Dissemination Co. Ltd. in Lhasa to do research and to develop Tibetan-style products from the rich Tibetan culture which are suitable for non-Tibetan customers. Every piece of the Tibetan craftwork demonstrates the Tibetan people's excellent traits of benevolence, sincerity, diligence and bravery. The products he sells, including silver accessories, basketwork and stone carvings, have unique designs and meanings and have attracted numerous buyers.

In 2007, thanks to the opening of the Qinghai-Tibet Railway which reaches Lhasa directly, the time it takes from Lhasa to Wuhan by train has been shortened to three days and the freight charge has been lowered to only 6.8 yuan per kilo. The rate of damage for products sent by air used to be 35 percent or so. But now, being able to send products by train has substantially reduced the rate of damage. To his great joy, after having visited Tibet, people from the interior become more interested in the Tibetan craftwork in which they used to see no value in use, while Tenzin and his team have kept renewing their ideas. The railway line has helped bridge the cultural gap between Tibet and other parts of China.

Tenzin eventually settled down in Wuhan. Lately, he often receives *zanba* (a Tibetan food, made of roasted barley flour) posted from Tibet. When he was little, Tenzin's family ate only two meals a day and *zanba* was their staple food. After living away from Tibet, Tenzin seldom ate *zanba*. He says that now *zanba* can be delivered so quickly by train that they still seem warm when they arrive. Last year, Tenzin received several Tibetan relatives at home who came to visit him by train. On every occasion, Tenzin asked his 12-year-old daughter to perform Tibetan dances for the elderly.

I Have Seen a Green China

| Liang Congjie | 1978: Editor of Encyclopedia of China Publishing House |
| | 2008: Retired |

In reviewing the past 30 years of his life, Liang Congjie regards the 10-year period from 1994 to 2004 as his most meaningful years, during which he and the Friends of Nature (FON) consorted with a good number of friends sharing a common goal and contributed their bit as a non-government force to the cause of environmental protection in China.

The scene of the then US President Bill Clinton having discussions with non-government personnel engaged in environmental protection in Guilin, Guangxi during his visit to China in 1998 remains fresh in his memory. Liang was lucky to be present at the meeting. When he presented photos of the rare and endangered Yunnan golden monkeys endemic to China to Mr. Clinton as a gift, he was asked how many of that species were still alive. Liang replied "no more than 1,200" and added that, apart from humans, they were the only primates with red lips. Mr. Clinton, looking at the photos with interest, exclaimed humorously, "Oh, they are my cousins."

The first half of Liang's life was closely linked to history study. In 1958, after completing a post-graduate program at the History Department of Peking University, he went on to teach history at Yunnan University. Later he worked for the Beijing Institute of International Relations, the Encyclopedia of China Publishing House, the International Academy for Chinese Culture and other organizations. It so happened that environmental issues caught his attention. In the early 1980s, Chinese township enterprises were booming and the news coverage of the mainstream media was filled with stories about how the township enterprises had promoted China's economic development. Liang was working for the *Encyclopedic Knowledge* magazine and received quantities of contributions every day. One day, he came across a particularly insightful manuscript, which forecast the potential damage that small and primitive township enterprises could cause to the environment. Enlightened by that manuscript, he began, for the first time, to be concerned with environmental issues. From that moment on, he has been thinking about contributing his bit as a common citizen to environmental protection. One day in 1993, he was chatting with a few friends. When the topic turned to China's environment, they came up with the idea of establishing a Chinese NGO like Greenpeace International to take environmental protection actions and to urge the government and mobilize the common people to make efforts in this respect. They decided to give it a try.

The history of Chinese environmental-protection NGOs can be traced back to 1978 when the China Society for Environmental Sciences was established following a proposal made by government departments. As the policy of reform and opening-up was being launched, all social sectors in China began to follow international practices. In the 1990s, China's academic community introduced the concept of "civil society" from developed countries to China. As public awareness of environmental protection was awakening, NGOs in this area flourished.

Friends of Nature initiated by Liang Congjie was founded in March 1994. Soon after, FON became the first Asian NGO to be awarded the Asia Environment Prize co-sponsored by *Daily News* of Japan and *Chosun Ilbo* of the ROK. In the early days after the founding of FON, lacking both funding and space, Liang used his home as the office and meeting place of the organization. In the first one or two years, almost all activity notification letters and envelopes were handwritten by FON members and stamped and mailed at their own expense.

FON gradually became well known to environmental activists for its considerable efforts in this regard. In the spring of 1996, the local government of Dechen County, Yunnan Province planned to fell a tract of virgin forest as a way of solving its fiscal problems, placing the habitat of Yunnan golden monkeys in danger of being destroyed. Environmental activists in Yunnan called out for protecting the rare species and went to Beijing, seeking to deliver their petition to the Central Government calling for protection of the virgin forest and Yunnan golden monkeys. At a meeting of the Chinese People's Political Consultative Conference (CPPCC), Liang Congjie, as a CPPCC member, managed to hand the petition over to the Central Government. Meanwhile, another environmentalist filed the petition to the central leaders through his own channels. Before long, the event attracted coverage in several media and gained strong public support. Ultimately, the central leaders issued official documents prohibiting the felling of that tract of virgin forest so that the habitat of Yunnan golden monkeys could be preserved.

As the public awareness of environment gradually grows, environmental protection has aroused increasing social concern. Yang Xin, a member of the Yangtze River rafting expedition, discovered through years of experience gained from his expeditions that the eco-environment at the source of the Yangtze was rapidly deteriorating and that the wild Tibetan antelopes on the Qinghai-Tibet Plateau, in particular, were in danger of extinction. The Tibetan antelope, or chiru, had long been listed by international conventions as a species banned for trade. Nevertheless, their wool fabric, known as the "shahtoosh," sold well on the international market. Liang Congjie and other non-government environmentalists decided to do their best to protect Tibetan antelopes. The then UK Prime Minister Tony Blair paid a visit to China on October 6, 1998. Liang was invited to the meeting between Mr. Blair together with other Chinese non-government individuals. Therefore, he wrote a letter in the name of the FON Chairman soliciting Mr. Blair's help in stopping the illegal trade of Tibetan antelope wool fabric in the UK and handed it to Mr. Blair at the meeting. In his letter, Liang wrote: "I appeal to you to use your influence in the UK and other EU member states, and together with them, to prevent the extinction of such a rare species." Mr. Blair wrote back to him the very next day, saying he would send his request to the environmental administrations of the UK and the EU and expressing a hope that it would be possible to terminate this illegal trade. In China, growing public concern for the fate of Tibetan antelopes also prompted action on this problem. The State Forestry Administration soon carried out the "Hoh Xil Action No. 1" campaign against the poaching of Tibetan antelopes in Qinghai, Tibet and Xinjiang. Hoh Xil is the major habitat of Tibetan antelopes.

Liang Congjie believes that, as an environmental protection NGO, what the FON can do is to assist the government in popularizing the concept of environmental protection among the people and strengthening the public awareness of environment. The FON has devoted most of its activities to this goal. To implement its Green Hope Initiative, the FON dispatched volunteers to offer environmental education to the youth and to impress upon them the concept of environmental protection by involving them in interesting games. The FON also offers environmental education to primary and middle school teachers. After an environmental lesson given to the primary and middle school teachers of the Zhoushan Islands, FON members led the teachers in collecting garbage in the local town. After the lesson, one of the teachers who had participated, said to Liang Congjie,

"I thought our town was quite clean, but after collecting garbage with you, I realize that I was wrong." Liang was very glad to hear that their activity had yielded such positive results.

Liang and his wife have been living a very simple life. They use bicycles daily for transportation, as they are convenient, nonpolluting and good to health. Now they are able to spend most of their money on concerts, books, DVDs and CDs. After watching the movie *Titanic*, Liang said to his wife: "Our planet resembles the Titanic in a way, with everyone living his or her own way and experiencing the whole gamut of emotions of life in cabins of different classes. But few know that our ship is going to bump into an iceberg. After this happens, everyone has the same fate. Therefore, we human beings must make concerted efforts to protect our only home planet."

It is by no means easy for NGOs to promote environmental protection on their own without the strong backing of the government. At the present time, Chinese environmental NGOs have begun to participate in policy-making, decision-making and government actions on this issue. FON is often invited to seminars on the legislation and amendment of environmental laws and regulations, such as the Energy Law and Law on the Prevention and Control of Water Pollution. In 2005, Green Watershed, Green Home, FON and other NGOs put forward rational proposals concerning the cascade hydropower development of Nujiang River to the Central Government and local governments at all levels and finally reached the common understanding that the feasibility of developing Nujiang River should be fully proved. In the same year, the All-China Environment Federation solicited public opinions and suggestions throughout the country on China's 11th five-year plan for the environment. Over 4.7 million citizens offered 27 suggestions in nine areas which were greatly welcomed by the State Council, State Environmental Protection Administration (today's Ministry of Environmental Protection) and other ministries and commissions. At present, there are nearly 3,000 environmental NGOs in China. Some scholars hold that environmental protection has become the most active field of Chinese NGOs.

Since Beijing succeeded in its bid for the 2008 Olympic Games, Chinese people's awareness of environmental issues has reached new heights. Motor vehicle pollution draws more and more public attention. Through the nationwide publicity campaign by the governments at all levels and by NGOs to advocate environmental protection, the environmentally friendly concepts of "green commuting" and "driving less" are taking root in the hearts of the people in Beijing. On World Environment Day and China's No Car Day, the staff of many companies use the subway, buses and bicycles or walk to and from work instead of driving cars, and some companies provide public transportation IC cards to their staff as a way of encouraging green transportation. The "Green Olympics, Green Action" Promotion Team delivered 350 lectures to over 100,000 people in 2007, the year with the most lectures, widest scope and biggest influence since its founding. Over the years, Beijing's efforts to make the city greener have focused on tree planting. By the end of 2007, Beijing had fulfilled all of its seven promises regarding Green Olympics. In 2008, the FON initiated an activity named "Global Warming Threatens Everyone — Plant 2008 Trees for Green Olympics." The FON provided 2008 seedlings free of charge to the public to plant in the Longmenkou Village, Zhaitang Town, Mentougou District, encouraging more people to plant trees, reduce emissions and slow down global warming.

Liang feels gratified that more and more young people actively engage themselves in the cause of environmental protection. Although he is getting on in years, he has seen a green China.

Once a Doctor, Always a Doctor

Li Ling	1978: Assistant physician at the People's Hospital in Kaixian County, Sichuan Province
	2008: Chief physician at the People's Hospital in Kaixian County, Chongqing; adjunct professor at the medical college of the Hubei University for Nationalities

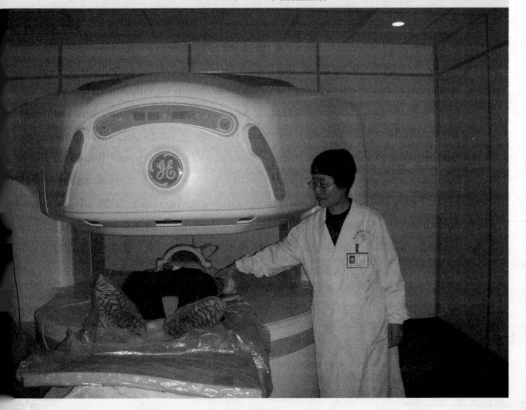

It was during the first year of China's reform and opening-up that Li Ling graduated from university. She would have stayed on at university, but she responded to the call to "put the focus of medical and health services on the grassroots units," and so she worked as a doctor in the hospital of her native county. And in the following 30 years, she continued to work in the hospital located in the north suburb of the county town. She worked there up until 2008, when the town was to be submerged as part of the reservoir formed by the Three Gorges Project. At this point, her hospital began to function in the new location of the town.

There were only 20-odd doctors in the hospital when she first arrived, with four clinical departments, including an outpatient service, internal medicine, infectious disease control and a surgical and gynecological department. Due to a shortage of doctors, Li Ling was required to be on duty alone after she had been working for no longer than a week. The first patient she treated was found with quick and irregular heartbeats, exophthalmos, thyroid gland enlargement and dysphoria. Based on her professional knowledge, she thought the patient should receive an examination of his basal metabolism to verify the syndrome of thyrotoxicosis. When she gave her opinion to an old doctor, he told her that only a few examinations referring to liver function, urinary routine and X-ray radiography could be implemented, while many other diseases could only be diagnosed and treated based on experience. Under the guidance of the old doctor, she treated the patient, who was discharged from the hospital a week later. After sending away the patient, Li Ling said to herself: although I studied for several years at the university, I had no means to say exactly the syndrome of the patient. She never thought that conditions in the grassroots hospital would be so inadequate. She realized that the only way to be a good doctor was to accumulate clinical experience.

Two years later, a graduate from Chongqing Medical University was assigned to Aunt Jiang's department. Aunt Jiang, who was Li's colleague and neighbor, felt that Li Ling had reached the marrying age, so she wanted to be Li Ling's matchmaker. Li Ling's mother, Old Tang, pretended to be a patient and went to the young man asking for an examination. She decided that the young man was handsome, honest and diligent, which was very much to her liking. So Old Tang selected a day when both her daughter and the young man were not working, cooked a large quantity of stuffed buns and invited the young man to her home to have supper together with her daughter. The young man had a good appetite and ate eight buns. After the supper, he helped her to clean up the house. Old Tang was very satisfied with him. Less than half a month later, the marriage was arranged. Neither of them could ask for a leave as the National Holiday was coming. So, they asked Aunt Jiang to take their place in getting the marriage certificate in the Bureau of Civil Affairs. When Aunt Jiang brought the certificate to the couple, she also brought them two one-yuan notes, which was the biggest cash gift that they received from their colleagues.

The wedding ceremony was held according to the customs of the husband's family. Her father-in-law, who had long been the secretary of the Party committee in his village, had a good relationship with the residents. The wedding procession queued for several hundred meters. What a bustling scene with much happiness and fireworks! Three years before, the young man had become the first university student from this mountain village, and then, he married his wife who was also a university student; besides, her father worked as an official in the county government. Many friends from neighboring villages came to congratulate them. Neighbors helped to arrange a magnificent banquet with more than 20 tables. Following local tradition, the guests played practical jokes on the newlyweds on the wedding night and after they left, her father-in-law and mother-in-law sent Li Ling 100 yuan as their betrothal gift. After three days of wedding leave, the couple returned to the hospital. They reciprocated their colleagues' gifts by distributing peanuts, melon seeds, cigarettes and fruit-flavored candies to each department.

In 1982, as in other parts of China, a group of young individuals with professional backgrounds were promoted to leading posts in her county. Li Ling was selected as the youngest member of the county Party committee, and was recommended as a candidate for a member of its standing committee. Her retired father resolutely opposed the idea of his daughter repeating his career and said to her: "You should steadfastly master the kind of skills required to become a good doctor to ease the pain of patients, and then you will be respected by people all your life." What her father suggested was what Li Ling was thinking, so she decided to be a doctor as before. After the county Party congress was over, Li Ling left for a year for advanced studies in Chongqing Medical University. When she left, her child was less than one year old and couldn't speak or walk. At that time, educational tapes for infants were not available, so she made her own tapes by reading fairytales and singing songs in Mandarin. . . Her husband played her recorded voice to their son every day. After she came back, her son hid behind his father and asked: "Who is that aunt?" The father told him: "That's your mama!" Her son refused her embrace until he recognized her voice the same as that played on the recorder. When the child called her "mama," tears came down Li Ling's face.

In 1984, Li Ling was appointed deputy-director of the county hospital, but she never left her clinical position. In 1998, she was made vice-chief of the county's health bureau. Li Ling's only request was that she be able to retain her license for practicing medicine. One year later, she returned to the hospital to resume her professional work because her mind was still with those patients who had gone to the health bureau for her help. Even at the present time, Li Ling's administrative position remains the same as it was 24 years ago, but her professional title has risen step by step from physician to chief physician.

In December 2003, an explosion that shocked the world occurred in Kaixian County. The natural gas field with a daily output of over 1,000,000 cubic meters ruptured, and the spread of high density hydrogen sulfide threatened the lives of 100,000 residents nearby. And the death toll increased sharply. At two o'clock in the morning, receiving the order from the rescue headquarters, Li Ling stayed up to work out the emergency plan. There was no experience to draw on because no grassroots hospitals across the country had ever encountered instances of mass poisoning on such a scale. She had to consult with chemistry teachers at a nearby high school as the library was shut down. Her husband helped her identify theoretical approaches from new university teaching materials. Thanks to her more than 20 years of accumulated experience, the emergency plan was completed and sent to the front line by five o'clock in the morning. Covering noses with a towel wetted with suds or clean water, tens of thousands of people escaped from the disaster areas and the hospital and clinics were swarmed with patients. When the drugstore ran out of supplies, she wrote a prescription for the pharmacy to make preparations to be sent to the health centers in disaster areas; when the wards ran out of beds, she made arrangements to set up beds in the county government conference hall and schools; when the medical staff couldn't cope, she gave some simple professional training to the caregivers in the hospital .... By the end of December, none of the patients who had been sent to the hospital for treatment had died. When a leader of the Central Government who came here on an inspection tour asked her what she needed, she said that the grassroots hospitals were very poorly equipped and that they should be improved to deal with public health emergencies.

A vice minister of the Ministry of Health was very moved and said it would be very lucky if there were a "Li Ling" in each county across the country. The Ministry of Health and the municipal government immediately decided to provide a special subsidy, and in the following several years, the investment in grassroots hospitals and emergency aid has been increased. In 2006, a flood disaster which had not occurred in over a century lashed the Kaixian County. This time, Li Ling was appointed general director of medical rescue. After the flood receded and patients were discharged from the hospital, Li Ling could finally stop worrying. The county seat was to be moved onto a new site on a higher ground five kilometers away. Such a flood disaster would never occur again because the new seat, built on a higher ground, is also protected by the Three Gorges Hydropower Station downstream which is capable of regulating flood waters.

The new county seat is the result of a state decision back in 1992 to build the Three Gorges Hydropower Station more than 200 kilometers downstream. The old county seat, together with two cities, another 18 county seats and 268 smaller towns, all lying below 175 meters above sea level, were to be submerged, and 1.13 million people were to be resettled. From 1996 onwards, they evacuated the reservoir zone in four major waves in what is the world's largest-scale organized relocation. Some villages and communities were resettled as a whole to developed areas to engage in farming work. Relatives of Li Ling's husband were resettled to Chongming Island in Shanghai; some were encouraged to pass college entrance exams or were employed in big cities. Li Ling's son was fortunate to receive very high scores during the college entrance exam and enrolled in a university in Beijing. More people, like Li Ling, moved to a new place nearby.

On January 1, 2008, Li Ling and her husband moved into a new house. Although the new house is situated in the downtown area, it is not as convenient to move about now as in the old town, which was smaller and people used to walk to any part. So, they bought a car. The new city is large, with broad streets, tall buildings and a clean environment. The quality of life has improved a lot. The biggest change Li Ling has observed is in the lifestyle, which includes using a vacuum cleaner when cleaning the house, taking plastic bags when walking a dog, waiting for the traffic lights to change before crossing the street and paying property management fees at the end of the month, and which had been unknown in the old town. A month later, Li Ling still found the new house strange and longed for her past leisurely life. Changes, especially changes in life, seem to be much harder for people over 50. In any case, Li Ling often praises the new city proudly to her friends from other places: "The new city is good. The Three Gorges construction has moved the county town up by 50 years."

On February 2, 2008, the People's Hospital of Kaixian County was set up in the new location. It was constructed according to the high standards of a third-class hospital with a floor space of 65,000 square meters on a plot of 118 *mu*, with 42 departments and sections and 750 beds. The hospital has been equipped with HIFU, Gamma Knife, magnetic resonance, multislice computed tomography (MSCT), bacteria-free laminar air operating room, ICU and KCU wards. Those medical services which were previously only offered in the provincial capital and Beijing can now be enjoyed in the county. On that very day, Li Ling and her husband were very happy and they had a picture taken together in front of the main building of the hospital to commemorate the great changes which have taken place over 30 years. But, what has not changed is that they're still doctors.

Dancing at the Intersection of Tourism and Aviation

Wang Zhenghua	1978: Vice-secretary of the Chinese Communist Party Committee of Zunyi Sub-District Office, Changning District, Shanghai
	2008: Chairman of Shanghai Spring International Travel Service Company and Spring Airlines Company Limited

From the late 1970s to the early 1980s, a great number of educated youths who had been sent to the countryside for reeducation during the Cultural Revolution returned to their urban homes and couldn't find suitable jobs. So, those young people waiting to be employed became somewhat of a social problem. At that time, Wang Zhenghua was the vice-secretary of the Chinese Communist Party Committee of Zunyi Sub-District Office of Changning District in Shanghai, in charge of economic work. In order to help those young people, he worked hard at identifying some promising industries, and finally resolved to launch a tourism company. The recruitment notice attracted more than 1,600 people who applied within two days. Wang Zhenghua selected over 30 of them for training. He invited some experts who had been engaged in the tourism industry in the past, even before the founding of New China in 1949. Those who had been recruited paid 40 yuan for the training. And this small amount of money became the total assets of the company at the beginning.

Launching a travel agency meant abandoning the "iron rice bowl" — his secure job as a civil servant. His colleagues tried to dissuade him from taking this step. His wife argued with him for nearly two months. Finally, one day, his wife came over to him and said: "Well, I have decided. Just do it. Whether you succeed or not, I will support you regardless of whether we only end up having pickles and congee for our daily meals." Even today, when he recalls his wife's remarks, Wang Zhenghua can't help getting excited. For nearly 30 years, as the chairman of an enterprise, Wang Zhenghua has always insisted on working on the front line. He gets up very early every morning while his wife is up even earlier in order to cook his breakfast, as this may be the only meal the couple can have together every day.

In the beginning of reform and opening-up, Chinese people were still comparatively poor, and few could spare money for travels. Before 1978, tourism in China was oriented to foreign tourists and there were no grounds for the development of domestic tourism. Shanghai was one of the cities that were at the forefront of the reform and opening-up. In early 1980s, there were group tours for those who went to Shanghai for conferences that took them to the suburbs and nearby cities. Travel agencies at that time focused on this kind of group tours while the number of individuals traveling at their own expenses was very limited. In the early 1980s, few people were researching tourism in China. In order to run his travel agency as well as possible, Wang Zhenghua would do research for relevant information from time to time. He was told that a tourism curriculum had been set up in Hangzhou University, so he asked one of his friends to bring him some mimeographed teaching materials translated by the university. He cherished them as treasures and read them day and night. Following the practices of travel enterprises in Europe and America described in the textbooks, Wang Zhenghua established a business strategy targeting at individual travelers, and began an intense search to identify places where people "with spare money, spare time and in leisurely mood" concentrated. After considering and reconsidering, he selected a commercial space near the busy Shanghai Great World Entertainment Center, a must-see place for visitors to Shanghai. A row of stores could form an aggregation of travel agencies convenient for potential travelers wanting to make comparisons. Practice has confirmed Wang Zhenghua's assumptions. This branch started to profit after only being open for a month, with the volume of business exceeding that at head-quarters. Actually, the so-called headquarters was just a rented two-square-meter pavilion in Zhongshan Park. Wang Zhenghua continued to stick to the principle he had learned from the teaching materials — "An enterprise should operate in accordance with the demands of the market." Frankly, the success of his company achieved over a decade should be credited to the edification provided by these teaching materials.

Since 1994, the Spring International Travel Service Company has ranked first in domestic tourism across the country. What Wang Zhenghua was considering next was how to expand business to new fields. Due to its status as a "private enterprise," the agency was only permitted to run a domestic tourism business aimed at Chinese citizens and barred from tourism business catering to foreign tourists and compatriots of Hong Kong, Macao and Taiwan. He wondered where there might be a broader market. Again, he began to research the development mode of foreign travel agencies which had combined the concept of tourism with aviation. In 1997, he took the initial step by contracting the one-year-long management rights of a plane. The costs of his travel agency were sharply reduced and the arrangement of flights became more convenient after he solved the problem of supplying airline tickets, a general choking point for domestic travel agencies. From 1997 to 2004, he contracted management rights for nearly 30,000 flights. In 2004, the General Administration of Civil Aviation of China (CAAC) resolved to break the monopoly in the aviation industry and announced that it would allow private enterprises to run aviation companies. Hearing this good news, Wang Zhenghua became even more ambitious and made further plans to launch his low-cost aviation company.

In China, there are lots of aviation companies running travel agencies but not vice versa. This is because travel agencies, with their slight profits, can't bear large investments and take high risks. At the meeting of the board of directors, when Wang Zhenghua raised his proposal to launch an aviation company, he tried to persuade the board members by presenting various reasons. However, they unanimously rejected his proposal. As chairman of the company, Wang Zhenghua, for the first time, decided to overrule the decision of the board. He believed that eighty to ninety percent of those traveling at their own expenses would like to be able to take a safe and inexpensive plane. He believed that low-cost aviation combined with tourism would directly cater to the demands of China's market. He also believed that this market opportunity would soon evaporate if he couldn't make good use of the policy aimed at breaking off the monopoly promulgated by the CAAC. On July 18, 2005, the first plane attached to Spring Airlines took off into the sky, penetrating the thick fog of the traditional aviation system. On that very day, the National Tourism Administration listed the Spring International Travel Service Company as one of the nine biggest tourism corporations in China. Since joining the WTO, the state's attitude towards private travel agencies has changed. At that time, the Spring International Travel Service Company not only operated inbound business but was granted special permission to operate outbound business aimed at Chinese citizens. Wang Zhenghua regarded these business scopes as landmarks in the development of the Spring International Travel Service Company.

It was his frugality, not his affluence, that helped Wang Zhenghua launch his aviation company. When he ran his travel agency in its early stages, he was saving all of the money earned with the exception of daily operating expenses and pay for manpower. He opened a new office whenever he felt he had saved enough money. From the 2-square-meter headquarters to 50 chain establishments in Shanghai, the Spring

International Travel Service Company also owns 34 wholly-owned subsidiaries within the country and seven abroad. Once, when Wang Zhenghua was invited to visit British Airways, the colleagues accompanying him suggested to him that he should dress decently to avoid any embarrassment. Wang Zhenghua insisted on wearing ordinary clothes and took the metro with his colleagues. When night fell, they couldn't find a place to rest, so they stayed in the basement of a hotel run by an Indian. There were no desks or chairs in the basement, so Wang Zhenghua, sitting on his suitcase, contacted his company by using his PC.

The aviation company launched by Wang Zhenghua, a thrifty businessman, targets those thrifty people who want to save their money. For the purpose of enabling more and more ordinary people to travel by plane, Wang Zhenghua has implemented his views on thrift to extremes. In order to reduce parking fares, airplanes attached to Spring Airlines try to steer clear of luxurious parking aprons and choose gate positions far away from the terminal building; Spring Airlines even rents some run-down and low-cost terminal buildings.

In order to avoid wasting food, there is no food supply on board; to save fuel, the company encourages customers to use carry-on bags; to reduce the cost of producing tickets, the bargain price ticket is available only through online payment and message information; in order to save resources, Wang Zhenghua sets the example first by using both sides of each piece of A4 paper in his office. All he does is purely for the purpose of lowering the price of the ticket. However, good intentions are hard to bring to fruition, and biases and questions have never stopped. Some people have complained that the delays are too long, some have complained that the service on board is unsatisfactory and some have even asked for compensation or have refused to get off the plane. Traditional aviation companies couldn't bear the "disturbance" brought about by Spring Airlines' cheap tickets. As a result, Wang Zhenghua became a target of public criticism. Meanwhile, Spring Airlines ran into a deficit in December 2005, but Wang Zhenghua still defiantly clung to his beliefs that ordinary people should be able to afford to travel by plane.

Wang Zhenghua opened a blog on the Internet in order to dispel any misunderstandings among the public. Atypical for someone who is famous, the articles on his blog were written by himself. He made full use of this kind of cheap Internet platform to publicize his low-cost aviation ideas as well as to resolve various complaints and criticisms he had encountered. The Spring Airlines completed its difficult take-offs again and again pushed by Wang Zhenghua's efforts. As Wang Zhenghua was getting old, he hoped that his son would eventually be able to help him run his aviation industry.

Wang Zhenghua hoped that his son would study abroad and learn from Western-style life and methods of work. But his son showed no interest in going abroad. Wang Zhenghua persisted in pressing his views and ultimately his son promised him that he would go to America. When he applied for a visa, the visa officer asked him why he wanted to go to America. He replied that it was his father who was forcing him to go. This reply made the officer burst out laughing, and the son successfully got a visa, something that was very hard to come by. In order to toughen his son, Wang Zhenghua gave him only 20,000 yuan. In America, his son was employed by a compatriot from Wenzhou and he also did part-time work in a casino. He earned two Master's degrees in seven years, and brought back 20,000 U.S. dollars. As a result, Wang Zhenghua felt assured and employed his son as the vice-president of the company.

Wang Zhenghua formed the habit of doing shadow boxing when he was young, and never abandoned it over the next 30 years. Wang

Zhenghua feels that this kind of exercise which stresses flexibility over strength helps him not only exercise his body but also comprehend the meaning of life. No matter how busy he may be, Wang Zhenghua always takes time out to do shadow boxing in the morning. He says that he feels that time is passing quickly and he is an old man now. Spring Airlines has just been established, leaving many affairs for him to handle. So his health is important to him and he feels that he must do shadow boxing no matter how busy he may be.

A Life Devoted to the Teapot

Sha Zhiming | 1978: Serving jail term
| 2008: Retired

On January 1, 2003, the first garden-like boccaro teapot museum opened in the ancient city of Nanjing. Located at 19 Nanbuting, it boasted 182 pieces of boccaro treasure made since the Ming and Qing dynasties, including masterpieces of Shi Dabin, Chen Mingyuan, Shao Daheng, Yang Pengnian and Yu Guoliang. Some of them were peerless. As a result, the museum attracted numerous domestic and international visitors, scholars and leaders at various levels. These 1,000-odd pieces of boccaro teapots were owned by then 72-year-old folk collector Sha Zhiming.

Sha Zhiming was known to all the elderly residents of Nanjing. Early in the 1950s, he had already made a name for himself in sports circles. He began learning martial arts as a child, focusing on weightlifting. In 1959, he attended Jiangsu People's Sports Meet and won the heavyweight championship in weightlifting. Shortly after that, however, during the "Three Years of Natural Disasters," his training team was forced to break up. At that point, he returned to the neighborhood factory to become a worker. In 1970, he suffered an injustice and was thrown into jail. During his term of imprisonment, his five-year-old daughter drowned and his wife left him, too. The unjust verdict left him with a broken family. On January 23, 1979, Sha was rehabilitated. Coming out of prison, he was all alone in the world, burying himself in collecting boccaro wares.

Falling in the definition of pottery, boccaro ware is adaptable and has an intense grain. A baked boccaro teapot has a lot of gas holes, which is good for absorption. It can preserve tea leaves for a long time when used as a container to store them, and retains the aroma of tea when used as a teapot. Many intellectuals in Chinese history have participated in the designing and manufacturing of boccaro teapots. Combining modeling, poetry, calligraphy, painting, seal cutting and carving, a unique culture surrounding the boccaro teapot has gradually taken shape: practical, for appreciation and with a focus on collecting values.

China adopted the reform and opening-up policy between the late 1970s and early 1980s. At that time, a wave of boccaro teapot collecting was sweeping Hong Kong and Taiwan. In the unenlightened mainland, however, many people didn't understand the colleting value of the boccaro teapot. Sha Zhiming is a Hui. The Hui Chinese like drinking tea because they eat a lot of beef and mutton. Influenced by his family since childhood, Sha loves drinking tea and therefore loves boccaro teapots. Coming from a well-to-do family and having many friends, he had collected a series of top-quality boccaro wares before the 1960s. Unfortunately, his collection was damaged during his term of imprisonment. In 1979, the neighborhood factory sent him to do marketing. He traveled around the country on business, which was advantageous for collecting boccaro wares.

Once he was on a business trip in Shanxi Province to buy fittings for the factory. He happened to hear that there was a boccaro teapot in Yuci. He visited the owner and, as expected, found a blue glazed teapot from the Qing Dynasty. Poems were written in white glaze on the teapot's body. It was an excellent example of boccaro ware, superbly crafted and with graceful handwriting. The owner was charging 6,000 yuan for it. At that time, Sha's monthly salary was only 48 yuan. After some bargaining, the price was finally dropped to 3,000 yuan. Since Sha hadn't brought enough money with him, he left a down payment and returned to Nanjing. Selling part of his ancestral house property, he went northward again to buy that teapot.

Sha's story of selling off his family's property to buy a boccaro teapot quickly spread in boccaro collecting circles. Then, someone told him that there was a royal teapot owned by a descendant of Qing-dynasty nobility in Northeast China. Hearing this news, Sha immediately left for the northeast. This time, he found a black teapot with a dragon pattern, made by Qing-dynasty master Zhu Fangxi. On the black body of the teapot was engraved a flying dragon bending down from the clouds, baring fangs and brandishing claws. At the bottom of the teapot were engraved patterns of sea and mountain symbolizing happiness as boundless as the sea and longevity comparable to that of the hills. The mouth of the teapot was decorated with a yellow glazed *ruyi*. Very powerful in appearance, it was a fine work even among the articles of tribute. This kind of color glazed teapot had been manufactured at a low temperature and in limited quantities. Therefore, such teapots are seldom seen nowadays. Securing this treasure, Sha was wild with joy. He held it in his arms and went back to Nanjing by train. Since the railway transportation system was fairly backward at that time, with fewer trains and running at slower speeds, the railway carriage was crowded with passengers. Sha was worried that his teapot might be damaged. Consequently, he pulled off his cotton-padded clothes to wrap the teapot, and held it in his arms throughout the journey. It was November and it was freezing cold in the northeast. As a result, he caught cold, giving rise to cholecystitis. After returning to Nanjing, he was laid up for a while. His son said disapprovingly of him for taking the teapot more seriously than his own life.

This folk collector searched high and low for valuable boccaro wares, visiting antique shops, peddlers of cultural relics, mobile stands and ancient big mansions. While saving money on food and expenses, he emptied his purse to collect boccaro wares. His aim was to obtain and pass on boccaro wares, instead of to make money. So far, he has not sold any piece to others. A famous painter in Shanghai was also a boccaro collector. Once he saw one piece in Sha's collection, a teapot in the shape of an eggplant, and immediately fell in love with it. So, he offered to barter two of his works in exchange for the teapot. However, Sha gently declined. Ya Ming, renowned painter and associate dean of Jiangsu Academy of Traditional Chinese Painting, often discussed teapot art with Sha Zhiming. He called Sha a "man infatuated with teapots," and the nickname spread fast. Sha Zhiming looked upon the boccaro teapot as his wife, son, teacher and friend.

After many years of collecting, Sha's collection continued to increase day by day. Moreover, he became a connoisseur and expert in boccaro wares. Aiming to carry forward traditional culture, display the quintessence of boccaro art and introduce his own collecting experience, Sha wrote special columns in newspapers and magazines, compiled monographs, gave lectures at universities, disseminating boccaro collecting knowledge by various means. His book *China Boccaro* was published in March 1996, and *Top-quality Ming and Qing Boccaro Teapots* edited by him was published in December 1997, with illustrations in Chinese, English and Japanese, providing a model for the study of boccaro culture.

Different from those collectors who kept their collections secret, Sha Zhiming wanted to make his collections public, thus displaying the culture and beauty of boccaro wares to more people. In the 1990s, the house in which he lived was rather small. However, Sha set up a shelf in the biggest bedroom to exhibit boccaro teapots, and slept by himself in the narrow hallway. His family complained about this, while he found pleasure in it. His home was a showroom as well as a place of research, welcoming many guests of taste.

In 1989, the Nanjiang City Government decided to establish the Nanjing Folk Custom Museum in order to protect cultural relics and develop research on traditional culture. In 1992, the Nanjing Folk Custom Museum opened at 19 Nanbuting, which was the former residence of the famous scholar Gan Xi. It had over 300 rooms. In 2002, it was completely restored and resumed its original form. At the invitation of the government, Sha Zhiming exhibited his collections in the Nanjing Folk Custom Museum. Entering the engraved brick door of the former residence of Gan Xi, crossing delicate courtyards, visitors would arrive at the richly ornamented boccaro museum. Boccaro wares, whose simple shapes matched the ancient house so well that they formed a perfect picture, were arranged in the cabinet in graceful disorder. Sha often received tourists and scholars from home and abroad in the museum, relating the history of boccaro wares and stories of man and teapot. In 2004, a Frenchman named Patrice traveled a long distance to the museum. He was also a boccaro collector and had bought many exported boccaro teapots. Patrice admired Sha's collection so much that, after returning to his own country, he sent Sha his book *Boccaro Teapots Exported to Europe*.

Sha has a happy old age. In 1988, he retired and then remarried in 1989. His new wife He Xiuzhen was also a worker. The old couple is dependent on each other and happy together. His children have all grown up, all with successful careers. His son, Sha Xiaoming, runs a printing company and his daughter is also in business. Sha Zhiming receives a pension of 1,000 yuan every month, in addition to pocket money of tens of thousands of yuan from his children annually. Sha lives a leisurely and carefree life in retirement. He starts a day with tea. Every week he takes part in activities of the Society of Amateur Performers of Peking Opera. He also practices his handwriting and visits friends, enjoying an interesting and full life. As a result, he is very healthy. Though Sha Zhiming is 77 yeas old, he still walks vigorously, and is quick-witted with a good sense of humor.

Though he lives a comfortable life, he never gives up his ambition of carrying forward boccaro culture. Although he is retired, he holds many social positions, including adviser in boccaro art of the Nanjing Folk Custom Museum, part-time professor of the College of Humanities & Social Sciences of Nanjing Agricultural University, member of the China Research Institute of Ancient Porcelain, member of the China Association of Collectors and adviser to Jiangsu Association of Collectors.

According to Sha Zhiming, a true life is one filled with useful activities and good deeds.

From Successful Contracts to a Placid Life

Ma Shengli	1978: Technician at Shijiazhuang Paper Mill, Hebei Province
	2008: Retired from former Shijiazhuang Paper Mill

The state-run Shijiazhuang Paper Mill had come to the brink of bankruptcy in 1984 and planned to contract itself out to a capable person who could turn in 170,000 yuan in annual profits to the mill. At that time, China's reform gradually shifted its focus from the countryside to the cities and was taking place in all sectors, including industry, commerce, science and technology, and education.

Earning 170,000 yuan in annual profits was by no means easy for a state-run mill in deficit. For some time, there was no response to the effort to recruit such an individual. The then sales director Ma Shengli finally stood up and said, "Isn't there anyone who dares to sign the 170,000 yuan contract? I do. Moreover, I will ensure 700,000 yuan in profits and double the pay of our workers in my first year of the contract. Should I fail to do so, I would have no complaints about any legal ramifications. " An annual turnover of 700,000 yuan to the state was an enormous figure. Some took it as a joke and most considered that Ma Shengli was simply bragging.

After Ma eventually took on the contract with the Shijiazhuang Paper Mill, he erected a 1.5-meter-high board at the gate of the mill, inscribed with the words "Mill Manager Ma Shengli." Ma fulfilled his promise by earning 1.4 million yuan in profits in the first year of his contract. This created a media sensation in the early years of China's reform and opening-up. The Xinhua News Agency's report "Good Mill Manager Ma Shengli Always Bearing in Mind the Interests of the Country and People" was carried in the major newspapers of the Communist Party of China. Vigorous activities relating to "Learning from Comrade

Ma Shengli" were thus unfolded in Hebei and neighboring provinces. His advanced concept of "breaking the system of lifelong employment and fixed pay" and the pioneering practice of "multi-level contracts and a clear definition of responsibilities for everyone" were new and fresh. There had been the cases of Bu Xinsheng "setting no ceiling or floor for pay" and Lu Guanqiu experimenting with employee share ownership prior to Ma Shengli. Although Ma was not the first to take on a contract for an enterprise in China, he was the first contractor to attain an annual profit of 1.4 million yuan.

In the past, state-run enterprises practiced unified Party leadership, which was later altered to the system of having the factory manager assume responsibility under the leadership of the Party committee. After the separation of Party and government functions, the system of having the factory manager assume overall responsibility was adopted and the Party committee became a supervisory organ. In December 1986, the State Council issued a notice about promoting a diversified system of contract management to give the contractors sufficient autonomy in management. This was interpreted as a policy of approval and encouragement from China's central decision-making authority toward contracting out state-run enterprises. During the second enterprise contracting boom launched in the latter half of 1987, 78 percent of all state enterprises under the state budget exercised contract system. At that time, there was a popular saying, "One capable person can save a factory." Under the contract system, managers played a more important role in running their enterprises. Ma Shengli had confidence in his ability to save 100 factories other than the Shijiazhuang Paper Mill. He began to set his sights on the entire country and mass-contract paper-making enterprises in 20 provinces of China. He planned to apply the successful model of the Shijiazhuang Paper Mill to another 100 similar enterprises.

At the height of his splendor, Ma was on a nationwide speaking tour giving over 600 reports on his glorious experience of saving and reviving every factory for which he had contracted. An article describes the scene of Ma giving a report in the following words: "He talked with wit and humor. The audience was quiet and intoxicated by his words. No one actually moved during his three-hour report and some even ignored calls of nature to the restroom for fear of missing any of his words." Ma Shengli was also a man of action. He signed dozens of contract agreements in succession, some of which were concluded after only one hour of negotiation. Ma's lightning successes intensified his image as a man of undoubted capabilities. All kinds of honors and titles were bestowed upon him one after another, such as delegate to the 13th National Congress of the Communist Party of China, National Model Worker, National Excellent Entrepreneur, and Young and Middle-aged Experts with Outstanding Contributions to the Country. In particular, he was the only person winning twice the National May 1st Labor Medal at that time.

On January 19, 1988, China Ma Shengli Paper-making Group was founded, one of the few privately-run groups with "China" in its name. The local mayor spoke at the founding ceremony, "Ma Shengli is a man, not a god." Ma's group managed to realize his plan of contracting 100 more paper-making enterprises, with Ma Shengli as the only legal representative. With satisfactory profits, he succeeded in creating a Chinese paper-making trust. That year was the first time that the title National Excellent Entrepreneur was conferred and Ma Shengli was one of the winners. The media commented that the issuance of the title marked the birth of the concept of "entrepreneur" in China and that the 20 winners that year made up a pioneer team in China's reform movement.

However, the good times did not last long. On July 30, 1988,

Guiyang Daily devoted a large space on its front page to the news report "Ma Shengli Having a Hard Time Managing His Contracted Guiyang Paper Mill." Ma, a high-profile individual, was infuriated by the report and denounced it as untruthful, one-sided, and obstructive to deepening the reform. Furthermore, he suddenly announced the termination of his contract with Guiyang Paper Mill for the alleged reason that *Guiyang Daily* had damaged his reputation and affected his work. A month later, *People's Daily* published a signed article, entitled "Some Thoughts on Ma Shengli's Setback." In the following day, the enraged Ma wrote an extremely sharply-worded resignation statement. Three days later, Ma Shengli arrived in Beijing alone and raced around in an agitated manner to over 20 state organs and news organizations, including the State Economic Commission, State Commission for Economic Restructuring, *People's Daily*, and *Workers' Daily*, to distribute his resignation statement. In early September of that year, several newspapers reported the news about Ma Shengli's attempt to resign and carried the full text of his resignation statement. Thereafter, Ma lost in the election of deputy to the People's Congress of Shijiazhuang and was no longer seen on the excellent entrepreneur commendation list of the city government. He fell as fast as he rose.

In May 1991, the Ma Shengli Paper-making Group was disbanded. The insolvent Shijiazhuang Paper Mill filed for bankruptcy in 1995, at which time the 56-year-old Ma Shengli, still four years away from the legal retirement age, was dismissed from office and ordered to retire. In 1997, the bankrupt Shijiazhuang Paper Mill and its 864 employees were taken over by the Chaoyang Enterprise Group. After that time, during the successive transfers in ownership of the paper mill, some of the original 864 employees were terminated with severance pay and some others had to negotiate the terms of their termination of employment contracts at the end of each year, with only over 100 still working at the original workshop of the paper mill.

In 2004, Feng Gensheng, one of the 20 winners of the first National Excellent Entrepreneur title in 1988, invited the other 19 winners to the West Lake Forum in Hangzhou, Zhejiang Province. However, only 10 of them participated, among whom only Feng Gensheng and Wang Hai, President of Qingdao Doublestar Group, were still holding enterprise leading posts. The organizer of the forum played a short documentary for those former pioneers of enterprise reform. The documentary mentioned that Ma Shengli had been forced to retire at the prime of his career and that the once powerful businessman had been given a retirement pension of only 135 yuan a month after his dismissal. Ma, a man of strong emotions, who for a time had to sell steamed stuffed buns for a living, burst into tears when reminded of his dramatic fall. At the forum, Wang Hai invited Ma Shengli to reenter business and take the posts of deputy general manger of Doublestar Group and general manager of Doublestar Ma Shengli Paper Co., Ltd.

After 10 glorious years followed by 10 secluded years, Ma Shengli resumed paper mill contracting. As times had changed, however, the contracting mode of "profit sharing and loss exemption" and pursuing short-term benefits, at which he had proven to be adept in the past, was already outdated. The modern enterprise system and share-holding capital reorganization became the focus in succession. The 65-year-old Ma Shengli wished that he could reproduce the past glory in his career while the Doublestar Group famous for its sports shoes hoped to enter the paper-making industry with the aid of the former reform hero so that it could expand into national manufacturing with an annual income topping 10 billion yuan in two to three years. Through several rounds of discussions, the Doublestar and Ma Shengli finally agreed that the Doublestar would provide capital and would contract the paper mill to Ma Shengli, who would exercise independent management over personnel, finance, property, production, supply and marketing, and would assume sole responsibility for its profits and losses. Ma could have his share of the profits under the premise of value retaining and increment of assets, but he would be dismissed in the case of losses.

This contract had a different background and implications from those of 20 years ago. This time Ma Shengli was faced with an entirely different market environment and more risks. When he had made profits for his contracted paper mills 20 years ago, he was only rewarded a pile of honorary diplomas and a salary corresponding to his managerial position. As regards the current contract, he could receive 50 million yuan for every 100 million yuan he would make in profits. If Ma succeeded, he would thus become a man of wealth and would not need to worry about his livelihood as he had been forced to do 10 years ago when he was dismissed and his managerial salary and allowances from the State Council were revoked. Nonetheless, the consequences of failure would be equally enormous. When his paper-making group went bankrupt, the losses were paid for by the state. Regarding the contract with Doublestar, he used his house evaluated at 300,000 yuan as collateral. His family would be homeless if he failed. With full confidence, Ma Shengli promised the media that he would make the best use of the last opportunity in his life and would become a multi-millionaire in five years. However, his contract with Doublestar soon came to an end for unknown reasons. But one thing was for sure, the background of this period was indeed different from before.

In his memoir *Life of Ma Shengli*, he ascribed his failure to his inclination to try to be a hero. Today, Ma is living in a neighborhood which was completed in 1997 in the Lixin community, Shijiazhuang, Hebei Province. Plenty of businessmen from other places live here and many of them deal in Doublestar shoes. The once famous Ma Shengli is known to all in the neighborhood.

Freely Travelling Between Eastern and Western Music

Tan Dun | 1978: Enrolled in the Central Conservatory of Music
from the Hunan Peking Opera Troupe
2008: A world famous composer

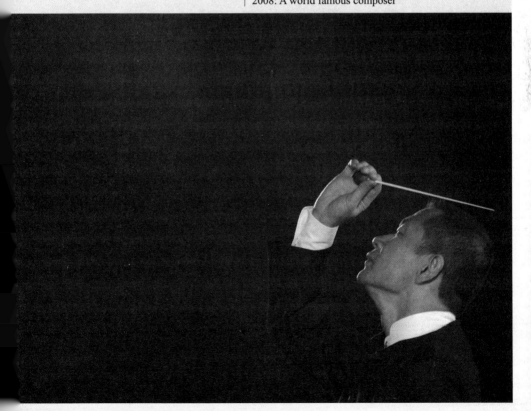

Tan Dun walked into the examination room of the Central Conservatory of Music carrying his violin which had only three strings.

In 1978, China reinstated the university entrance examination. At that time, Tan Dun was a musical instrument player with the Hunan Peking Opera Troupe. He decided to change his own fate through study. He worked very hard for the examination. He often listened to music that was being broadcast and wrote down the music scores under the big tree outside his house. At the time, few music books were sold at Xinhua Bookstore, not to mention violin music scores. Aiming to learn more compositions, he had to borrow music scores from others, and spent the whole night copying them. Every morning, he climbed to the top of a nearby hill to play his violin based on the music scores, often resulting in a sore hand and a shaky arm. Under the scorching sun, he practised violin, with his bare arms dripping with sweat. His family had no electric fans at that time. So, his mother silently kept fanning him from behind.

During the examination, he was asked to play a famous violin melody. He, however, played a melody which he had himself composed based on Hunan folk music. Benefiting from this melody, he passed the entrance examination of the Central Conservatory of Music.

In the early 1980s, the art world was rebelling against the art concepts of the Cultural Revolution, and turned to focus on human nature and factuality. The atmosphere of artistic creation became relaxed in an unprecedented manner. Tan Dun, who had always considered himself brave and unrestrained in the pursuit of art, and an elite of his era, felt himself bound to participate in the great project to reconstruct ideal and culture after the Cultural Revolution. During the summer vacation of his sophomore year, Tan Dun composed his first symphony in the classroom — *Li Sao*. He used many folk instruments including the *xiao* (a vertical bamboo flute) in this work to interpret the poem of the same title (meaning lament or sorrow after departure) by the great poet, Qu Yuan, who lived more than 2,000 years ago. At the time, this was going a bit too far. His teacher was not satisfied with his symphony, and asked him, "Do you really have such deep thoughts? Do you really have so many complaints?" Although his symphony evoked much controversy in the art world for using modern sound equipment and technology, he was extolled as one of the four top talents of the Central Conservatory of Music by some teachers and students who approved of his symphony.

Tan Dun worked unflaggingly for artistic creation, unconcerned with the comments of others. In 1983, he composed his first string quartet, *Feng, Ya, Song* (three parts that make up *The Book of Songs*), widely incorporating folk tunes. It gave foreigners a fresh and novel impression, thus winning him the Weber Prize in Dresden. In 1984, several melodies were presented at Tan Dun's Concert of Chinese Instrumental Music, evoking much controversy and shaking the world of folk music.

Recalling his nine years in Beijing, Tan Dun described himself as an arrogant youth who was thirsty for knowledge. At the time, he was very young, full of vim and vigor. He believed modern art trends were to negate all before bringing forth the new. Therefore, he chose to focus on pioneer music and use instruments which he had made by himself. He even tried to produce sounds from human organs, including the nose, ear and throat. He worked hard to make his creation attain such a level that all the instrumental ensembles could only be performed by his band and all the music scores could only be understood by him.

Tan Dun found that Beijing could no longer satisfy his desire for creation. So, after earning a master's degree in composition from the Central Conservatory of Music, he went to New York with the ambition of changing Western music, carrying with him a suitcase of toilet paper because someone had told him that it was very expensive there. In the United States, Tan Dun won a scholarship from Columbia University to learn music creation.

New York was crowded with various people, including such young artists as Ai Weiwei and Chen Kaige from Beijing, Chen Danqing and Chen Yifei from Shanghai, as well as Ang Lee from Taipei. Tan Dun lived in the artist-inhabited Greenwich Village for over ten years. During this period, he spent almost every day with Ai Weiwei. Though they were poor, they were ambitious. In Greenwich Village, Tan Dun learned how to "look at the world from the universe," and also understood the extensiveness and depth of Chinese culture. Therefore, he decided to combine Chinese factors with Western music. He was confident that he could change Western music and change the music of the world.

He claims to have spent ten years watching all the classical Western operas, many in several editions. After ten years of hard work, Tan Dun gradually came to understand what traditional Western audiences were like, who liked Shakespeare, what challenged Western opera and what they wanted to change. For this reason, he rarely published his works until 1989 when *Nine Songs* (inspired by Qu Yuan's poems of the same title) performed by self-made instruments was approved by the international musical world. In 1990, he tried to combine theatrical performance with band, and composed *Ghost Opera*, toured worldwide by the Kronos Quartet and *The Gate*, premiered by the NHK Symphony, earning a positive assessment from the Western musical world for their unique mode of expression.

In the meantime, musical trends were changing in China, too. The early 1980s were witness to a blind rebellion in the musical world. In the mid-1980s, art was not only a form of social criticism, but was also related to life. In the early 1990s, opposing the art idealism of the 1980s, the emphasis was laid on man's living conditions. In the mid-1990s, along with the influx of the Western consumer culture into China, Chinese modern musical trends and Western music were coordinated in their development. When something new appeared in the West, Chinese artists would soon follow it. Integration was the fashion of that period.

In 1995, Tan Dun was invited to write the music for the film *Nanjing Massacre*, directed by Wu Ziniu. In order to do it properly, he returned from the United States to observe and learn from real life. He visited the Nanjing Massacre Memorial Hall to review the relevant history, and devoted his heart and soul to the creation of this musical score. Different from his usual style, he composed a solemn and heroic song — *Don't Cry, Nanjing*.

Tan Dun showed originality in his understanding of Chinese culture, especially in his early years of his creativity. His inspiration mostly came from traditional Chinese culture and "mystic" culture. Moreover, the experience gained from studying abroad enabled him to use modern and post-modernistic music in the West for reference. Due to his special background, he was commissioned to write *Symphony 1997: Heaven Earth Mankind* for the official ceremony for the transfer of the sovereignty of Hong Kong to China in July, 1997, which was the only topic composition he had accepted so far. Different from his former pioneer style, it was standard in its musical form, tune and words. It abandoned cold-war thinking, avoided political words, and returned to a traditional mode. The symphony was based on Chinese culture and used Western musical form for reference, which was a compromise between the East and the West, both traditional and modern. Assembling folk songs and rhymes, orchestral music, symphonic music and a cello solo, it harmoniously mixed Western and Eastern music within a global scope. *Heaven Earth Mankind* was a cross-century work of Chinese classical music, marking the entry of Chinese artists into the international musical world.

In the same year, Tan Dun created the opera *Marco Polo* which arrested world attention then, and brought him much praise and many international prizes. His later works were all successfully performed, giving him a fast-growing reputation. In 2001, he spent 10 days in composing the song for Ang Lee's film, *Crouching Tiger, Hidden Dragon*, which won an Oscar for its original score. Since then, his name is known to more people outside of musical circles. He won a Grammy Award in 2002, and was honored by *Musical America* as "Composer of the Year" in 2003.

On December 12, 2006, *The First Emperor*, composed and conducted by Tan Dun was premiered at the Metropolitan Opera in New York City, with a title role created for Placido Domingo. It was the first Chinese-made opera to be presented at the prestigious Metropolitan Opera House. Nine performances were run in succession at the Metropolitan Opera House, and the crowd overflowed at each show even though the highest price of admission reached 2,500 US dollars. In the eyes of the foreign media, this opera combined the features of both Eastern drama and Western tradition, shaking the foundations of Western art. By the end of 2007, it had received invitations from 107 opera companies all over the world, which was a huge success from a business standpoint. At this point, Tan Dun believed he had realized his dream of changing

Western music. Yes, it really had "changed the mechanism of world opera," just as French and Italian operas once had.

Today, Tan Dun's works are met with a mixture of praise and stricture in China, just as his first symphony *Li Sao* was. Fortunately, art creation is booming in the Chinese musical world while art criticism has become diverse. The demand for art and art appreciation have become multi-faceted.

A Bus Rider's Six Moods over a 30-year Period

Wang Jianwen | 1978: Worked in Xi'an Tumen Photo Studio
2008: Retired from the Andingmen Photo Studio of Xi'an Xijing Color Photography Service Center

Xi'an has already had a tenfold increase in size since the beginning of reform and opening-up 30 years ago. However, Wang Jianwen, a 55-year-old woman who has spent most of her life in Xi'an, thinks that, despite the tremendous expansion of the city, today's omni-directional transportation networks and various vehicles make it more convenient for people to get around than it was in the past.

What made her proud: owning her first bicycle

As a child, Wang Jianwen moved to Xi'an with her parents who devoted themselves to the construction of the Northwest China. In 1978, when she was 25, Wang Jianwen worked in a photo studio in Tumen. Every day she had to go to some locations across the city to collect payments, so she bought a Phoenix bicycle subsidized by her work unit. She is still proud to recall those happy times when she rode her bicycle: "At that time, people couldn't go far, and there were few commodity networks. Most people would go out just to work, and only on foot or by bus. Riding a bike to work every day made me feel good."

In 1979, Xi'an had 498 buses running on 34 bus routes, covering 278 km. The daily number of passengers taking the bus for a home-to-office journey reached 350,000 person-times. Most people left home on foot. Only those who were doing well economically owned a bicycle.

What worried her: always having to get on the bus with a shoulder push from the driver

According to Wang Jianwen's recollections, in the 1980s the city was grey. The streets were bumpy and dimly lit and the telephone service was poor. But her worst recollection is that there were few buses and bus routes. Many places were inaccessible by bus. If you planned to visit your relatives, you had to ride your bike and carry all your family members. Besides, far too few outmoded and broken-down buses were crammed with far too many passengers. Furthermore, you had to wait at least ten minutes or sometimes even an hour for one bus.

"Actually, you couldn't get onto the bus in a normal fashion. What you had to do was to squeeze yourself onto the bus. If you made it, you were very lucky. Nowadays, passengers are so fastidious about the conditions on the bus. They can pick and choose a bus equipped with air-conditioning without the congestion of too many people, and use their trendy transport cards." Wang Jianwen indicated that what had left the greatest impression on her was the fact that some passengers, seizing their opportunity, tucked themselves as much as they could into the bus. Usually, bus drivers had to help passengers onto the bus with their shoulder pushes to ensure that the door could be closed to drive the bus away safely. Sometimes, drivers would feel sorry for the passengers and permit them to get on the bus through their driving cab. Catching a bus was tough work, so seizing a seat, which caused disorderliness and thefts on the bus, was a very common phenomenon.

"Although there were also some taxicabs on the road, most of them ran without passengers. Ordinary people just couldn't afford them." She added that some of those taking taxies were foreigners or rich businessmen from Hong Kong and Taiwan, and some of them were domestic businessmen from the southeastern coastal areas who had recently become rich in Xi'an as a result of their business successes.

At that time, the area accessible by buses in Xi'an was limited to within 90 square km. Meanwhile, at the beginning of the reform and opening-up, tides of peasant workers and students contributed to the acute strain on the means of transportation in the 1980s. When you went to work or went home, you had to wait for at least two buses before you could get on. At rush hour, ten people had to squeeze within one square meter, and the full-load reached 120 percent capacity.

In 1982, the Standing Committee of Xi'an City Party Committee set out to resolve this conundrum and set the goal that all passengers waiting for a bus needed to be picked up in two bus runs at rush hour and in one bus run at other times. The government supplied 100 buses each year for the next five consecutive years to maintain the rapid increase in the number of buses. By the end of 1989, there were over 800 buses with 54 bus routes in Xi'an.

What made her nervous: reckless minibuses

Wang Jianwen's impression is that, since the mid-1980s , the embarrassment involved in taking a bus has been eliminated.

The tempo of city construction in Xi'an was quickened after 1990. As the network of roads was gradually improved, citizens' requirements for enough bus routes were becoming more and more urgent. Due to the lack of financial support, the government, in order to deal with this conundrum, relaxed its restrictions on the public transportation industry and tried a market solution with the concept of "multi-ownership management, balanced development and unified administration."

Some "wave-and-stop" minibuses run by individuals began to appear in Xi'an. If you had an emergency, it was convenient to be able to wave to stop these minibuses moving quickly on the streets. However, they were much more expensive and not well regulated. There were also lots of minibus routes. Minibuses relieved the heavy burdens on public transportation but also caused potential safety hazards. Even now, Wang Jianwen can clearly remember how her colleague's feet were injured by the door when she got off a minibus. And once Wang Jianwen's bag was pinched by the door and she cried out for help, but the driver made no response. So, she determined not to take the "wave-to-stop" minibuses any more unless she had an emergency.

After five years of regulation effort by the government, minibuses gradually returned to orderly operation and, to some extent, became a form of transportation which was supplementary to public transportation.

What made her grateful: a warm-hearted taxi driver

China's taxi industry expanded rapidly in the 1990s. The number of taxicabs in Xi'an reached over 1,000 in 1995. There were 48 new taxi companies founded during 1993. The taxi company license was issued by public auction instead of administrative approval in the past. On August 23, 1995, the government auctioned off a ten-year-long management right of 600 taxies at a price of close to 38,000 yuan.

Wang Jianwen's gratitude to taxies was the result of generous assistance from a taxi driver to her niece. In 1995, her niece suffered from a blood disease. The only wish of the poor little girl who had never taken a taxi was to have a taxi ride in Xi'an before she died. At the front gate of the hospital, her family stopped a taxi and let the girl have a ride around Xi'an for the whole afternoon. During their tour, the taxi driver got to know the truth and he turned off the taximeter, telling the family that he couldn't do more to help the girl but would accept no payment for the trip. Finally, after the family's repeated requests, he took only a nominal 10 yuan and left. Later, the little girl miraculously recovered. And the family came to pay more attention to Xi'an's taxis just out of gratitude.

Meanwhile, the type of taxicab in Xi'an has undergone more than four stages, and the low-grade vehicles had already been phased out by the late 1980s. Even those red Xiali and Alto cars which had been popular in the 1990s were gradually becoming rare. Although relatively more upscale vehicles, like Fukang and Jetta , emerged in the third stage with a greater power and less pollution, and they still have a dominant role in the market, vehicles which emerged in the fourth stage in recent years, like Hong Qi, Passat, Lingyang Chery, Santana, Santana 3000 and BYD, etc. have made the drivers of old-fashioned cars very jealous. They had thought that maybe those newly added vehicles were so expensive that nobody would give them a try. However, those relatively more upscale vehicles have now already become the mainstream. Changes in the taxicab industry not only reflect rapid development of the city, but are indicative of the improvement in people's lives. Furthermore, they also show that the whole society has taken a great leap forward.

What made her love the public transportation system: it returned to the people

In 2007, the city government drafted and issued a strategy giving priority to the development of public transportation, and also launched the first "Public Transportation Week" and the first "Car Free Day" to provide citizens with more traffic resources.

Now, she thinks it's very convenient to travel by public transportation. You can find bus service wherever there is a road. Even if there is no bus, it's easy to find minibuses beside your home or to take a taxi. However, she seldom takes taxis. She always thinks that "taking a bus was like torture in the past, but now, it is an enjoyable thing." At present, there are enough buses available with comfortable conditions for riders and passenger-oriented services, and the trials and tribulations of trying to crowd onto the bus have passed. You never hear shouts from drivers hurrying you to get on or off the bus, and you can completely relax when you take a bus.

To her further delight, in 2007, passengers began to enjoy the preferential treatment of a fifty percent discount by paying with their IC card when taking a bus in Xi'an. "I envied my friends in other cities who were enjoying a reduced payment by using their IC cards when taking a bus. Now, we can also enjoy the same privilege in Xi'an in less than a year." She especially called her friends in other cities to say: "We only need to pay five mao (ten mao to one yuan) by using the IC card for a ticket whereas you need to pay seven or eight mao." She was told that only four mao are required in Beijing to take a bus and it only costs two yuan to travel on all subway lines.

Being a warm-hearted woman, Wang Jianwen began to keep a bus diary, in which she noted down which ones served passengers well and which ones had a good driver. She would then provide this feedback of all she had recorded to bus companies. At the beginning of this year, she became one of the first model passengers in Xi'an.

Currently, the bus passenger capacity approaches three million person-times per day, ten times the number 30 years ago. The number of bus routes is 204, almost octupling that of 30 years ago. And the number of buses is 6,000, 12 times as many as the city had 30 years ago.

What she most looks forward to doing: taking the subway

"I've taken the train and even a plane, but never the subway. I really hope the construction of the subway will be finished as soon as possible." When she talked about the subway which was under construction, she couldn't help yearning for its completion. In her opinion, a subway should be regarded as the symbol of modernization for a city, and other big cities like Beijing, Shanghai, and Guangzhou, etc, already have their own subways. Although there are numerous underground cultural relics in Xi'an, after a thorough review, the city government made the decision to build the subway. It's said that the subway will bring no harm to cultural relics, and it will be environmentally-friendly, highly-efficient and convenient. She says she can't wait for its completion any longer.

Wang Jianwen smiled and said: "Think about the future in which subways, buses, taxicabs and minibuses will make the out-going trip an easy thing for passengers. You will have various alternatives, like a train, car or plane to choose for traveling around the country or going abroad. By the way, it only takes about ten hours to go abroad by plane."

Nowadays, several dozen meters under this ancient city, the construction of Subway Line Two is in progress and another five lines are planned and will be built one by one. Maybe, within 50 years, six subway lines covering a distance of 251.8 km will become a new artery of this ancient city. These subways will reach as far as each main district and alley in the suburbs, offering services to 52 of the 61 main passenger flow corridors and large-scale passenger terminals with a daily carrying capacity of 5.095 million passengers. The subway will become most people's alternative and shoulder the traffic burden together with the bus, taxicab and minibus in Xi'an.

Finding a Village's Path to Wealth

Wu Renbao: | 1978: Secretary of Huaxi Village Party Committee and of Jiangyin County Committee, Jiangsu Province
2008: Vice-chairman of the board of the Huaxi Group and its deputy general manager, head of the General Administrative Office for Party, Village and Enterprise Affairs of Huaxi Village, chairman of China's Research Society on Prosperous Villages, and deputy chairman of China's Development Society for Poverty Reduction

"Without Wu Renbao, Huaxi Village would not have had what it has today," say Huaxi Village residents. "I owe everything I have today to Huaxi Village, " says Wu Renbao. Huaxi Village was formed in 1961, with Wu as its Party branch secretary. He served as head of Huaxi Village at 22; head of the township at 27; head of the county at 53; and again Huaxi Village head at 59.

Huaxi village had been poor in the past, with some extremely high and some extremely low ground, thus suffering drought after half a month without rain, and water logging in times of heavy rain. In the late 1970s, when China launched a campaign of cutting off the "capitalist tail," that is, any vestiges of non-public and non-farming economy, Wu believed firmly that there could be no stability without agriculture, and no affluence without industry. He led other villagers in starting a small hardware factory, making a profit of over 300,000 yuan that same year. Several years' accumulated profits transformed the villagers' houses into new homes with tile roofs, and each villager could be given 220 yuan instead of 130 yuan. The villagers also built a five-story building, locally called the "Education Building," for a nursery, and for an elementary and high school .

In the early 1980s, China launched major reforms in the countryside, popularizing its policy of establishing a household contract system. Wu knew that the policy was meant to make farmers rich. He reasoned that it was necessary to distribute land to each household in areas where villages had more land but fewer farmers, whose enthusiasm had long been dampened by "eating from the same big pot" (getting equal pay for unequal work). But in Huaxi Village, if all the land were to be distributed , there would only be a half *mu* (0.03 hectares) for each person. With a burgeoning population, the situation would likely become even worse in future. There was no way out if farmers were confined to the land. How could they escape poverty? Wu thought about this for a long time and concluded that the village should also develop industry.

The household contract system was promoted in the countryside, with only one voice heard: the contract system was a solution for every household. Whoever did not sign a contract was seen as disobeying the Central Government's policy. This put Wu under unimaginably great political pressure because the village kept to the practice of common prosperity through collective ownership. In May 1980, Wu, then Jiangyin County Party Committee Secretary, even lost election for a delegate to a forthcoming CPC congress of Jiangyin County. When it was subsequently decided that Wu would be transferred to head the Department of Agriculture and Industry at the prefectural level, he asked to return to Huaxi, saying that since he was a farmer he had only one wish, which was to work more for fellow farmers. As a result, Wu once again became a farmer in Huaxi. At a local Party branch committee meeting, he said, "The primary task for Huaxi is to develop the collective economy, letting everybody enjoy a wealthier life in every respect." He shifted most of the laborers to industry while assigning only 30 skilled villagers to farm 500 *mu* (33 hectares) of land.

In 1983, China initiated a system of direct elections for village heads. Raising the slogan of common prosperity, Wu stood for the election as Huaxi Village head. He explained his slogan in the simplest words: sharing a happy and wealthy life, without any families left behind. He was elected unanimously and remained the respected village head for a long time. In 1986, the total production of Huaxi village exceeded 100 million yuan. This sensational news meant that it could be held up as a successful example for other villages nationwide. Since then, over the past 20 years, Huaxi's economy has grown rapidly at an annual rate of 20 percent, with a strong momentum for future development. The village's fast growth has been attributed to the implementation of Wu's idea — less distribution and more accumulation; less cash distribution and more reinvestment as shares. He believes that it is important for rural enterprises to rely on self-accumulation for their development, rather than on government loans or funds collected by promising high interest. Only when adequate money is amassed can a virtuous economic cycle be ensured in Huaxi Village.

To suit the needs for opening up, the villagers, who had moved into new homes, built a 4,000-square-meter hotel on the site of their former residences. The hotel has three classical style corridors and its decor features simplity of southern China's rural areas. In the following year, a farmers' park with a lake and rockeries was constructed to the southwest of the village. All these developments demonstrate Huaxi Village's readiness to welcome tourists.

In 1988, the craze for establishing companies swept the country. As more people became focused only on money, Wu became very worried. He did not believe that money was all powerful. He was determined to advance ethical and material progress in his village. He created an unprecedented organization termed "ideological development corporation." He selected well-educated, respected Party members to manage the corporation. Whenever there were signs of gambling and superstition, the corporation would hold lectures, calling villagers' attention to the harms of gambling while explaining the difference between superstitution and religion; whenever there was family discord between husband and wife or mother and daughter-in-law, a short session would be organized to teach something about marriage and inheritance laws, with an aim of spreading traditional Chinese virtues such as respecting the old and caring for the young; to serve the village's economic interests, the corporation would hold lectures on enterprises, management and marketing. These lectures did not produce any profit on the surface; their benefit can never be measured in the sense of profit. The village benefited from tremendous improvement in its residents' moral standards and in the concerted efforts for better production. All of these are intangible assets for Huaxi Village.

In the 1990s, the country initiated the reforms relating to property rights for rural township enterprises. Huaxi Village did not blindly follow suit, but instead created its own innovative approach, namely "one village, two systems," whereby villagers could enjoy both collective and private ownership, while cadres could not. In the countryside, without a collective economy, the promise to help the masses is an empty one. The village has developed rapidly with the collective economy as its mainstay, supplemented by other economic elements. Today's Huaxi Group Company, the backbone of Huaxi's economy, deals in a wide range of businesses, including aluminum, steel and copper bars, textiles, the chemical industry and tourism. In 1999, the Group became a listed company in Shenzhen and it was also the first one to be named after a village.

Since the 15th CPC National Congress in 1997, the Central Government has made proposals for effectively managing large enterprises while relaxing controls over small ones. Wu knew a great deal about Huaxi enterprises. While following the Central Government's policy, he also suggested the idea of supporting small enterprises as well and assisting them as much as possible when needed. Huaxi's large enterprises, also the backbone of the local economy, including textiles, wire-making and steel bar factories, should be given more support in management and in reform. Some smaller ones with low profits should be turned into privately-owned enterprises. Those more promising smaller enterprises should be offered intense support in terms of funding, professional expertise and technology so as to enhance their competitiveness. The reforms in Huaxi enterprises have been carried out in an innovative way, thus greatly stimulating the enthusiasm for production. At the present time, its more than 50 enterprises have maintained an annual rate of growth in their profits, with none in the red. In 2007, the village yielded 45 billion yuan in sales, and turned in more than 800 million yuan in profits and taxes.

Economic development has brought affluence to Huaxi Village with the collective economy as its strongest feature. The villagers who are retired and disabled receive a monthly pension, while the others work in the village's enterprises, with their lunches covered. A villager is normally paid less than 1,000 yuan a month, but bonuses can be four times as much as that figure. With a full monthly salary, a villager is given only 20 percent of the bonus in cash in addition to some other dividends. In 1984, when color TV sets first appeared on the market, each household in this wealthy village was allocated one set. In 1992, the village spent two million yuan installing wall-mounted air conditioners for every family. In 1993, after telephones, color TV sets, air conditioners and villas were made available to every family, Wu and other village leaders bought cars from the Changchun First Automobile Works for their fellow villagers. After achieving affluence, the village has turned its attention to the environment. Now eco-tourism has become a highlight of Huaxi's economy, attracting over two million domestic and international tourists a year.

At the present time, an average family's assets in Huaxi Village amount to at least one million yuan. Nearly all the villagers live in fashionable villas, while Wu still lives in an old building built in the 1970s. His group photos on the walls, squeaky floors and old-style double bed have accompanied Wu and his wife for 30 years. At first, Wu said that he would move into a villa after all the other villagers did so. Now, he says he is used to his old home. For decadés, he has kept his word not to have the highest salary and bonuses, nor to live in the best house.

Wu has a large family made up of twenty-eight people of four generations who live happily together. His children all have aspirations. They once wished to create their own world outside the village. Wu wants them to stay in the village for Huaxi is what he is all about, urging them to hold onto their roots and contributing to their hometown.

Witnessing the Changes in Zhengzhou

Wang Shuguang | 1978: Doctor in the Worker's Hospital of the Academy of Agricultural Sciences in Henan Province
2008: Associate chief physician in the Worker's Hospital of the Academy of Agricultural Sciences in Henan Province

When she was a middle school student in Guangzhou, capital city of Guangdong Province, Wang Shuguang responded to her country's call to participate in reclaiming wastelands and guarding the frontier for the construction of Hainan Island. In 1973, as educated city youths who had been sent down to the countryside like her were returning to urban areas in large groups, Wang Shuguang also followed her parents who were transferred to Zhengzhou. Now, several decades have passed quickly, and when she recalled that exciting period in Hainan, she longed to step onto that land again. However, due to her busy schedule, this long-cherished wish hadn't yet come to fruition. On January 1, 2008, the state began to strictly carry out the Announcement of the State Council on the Regulations for Paid Annual Leave of Employees and some relevant departments issued more detailed implementation rules, guaranteeing all employees the basic rights of annual leave. After the adjustment for national public holidays, the average annual days of leave for all employees account for over one third of the whole year. Many colleagues have made full use of their leave traveling around the country, and Wang Shuguang also boarded plane to Sanya, a resort city of Hainan Island. As she flew near, Wang Shuguang was moved by what she saw through the window — it is now a modern city with numerous skyscrapers and crisscrossing highways. At this point her mind also flashed back to the changes that had taken place in Zhengzhou over the last 30 years.

In 1978, Wang Shuguang became a doctor in the Worker's Hospital of the Academy of Agricultural Sciences in Henan Province. At first, the hospital, with five medical staff members and three rooms, only had the means to diagnose ailments like colds, fever. Wang Shuguang was passionately devoted to her profession and always worked carefully. She won universal praise from the staff because she often provided medical treatment to mobility-handicapped patients after work or on holidays. After 2000, the state advocated the policy of "treating serious diseases in the hospital and handling minor ailments within the community" and allocated funds to establish community hospitals to make it convenient for community residents to obtain medicine or transfusions near homes, so they would no longer need to line up to register at the hospital. And the Worker's Hospital of the Academy of Agricultural Sciences also developed rapidly and was equipped with some new medical facilities. Compared with the past, levels of service and professionalism have been greatly improved. In recent years, Wang Shuguang's colleagues have been trying to upgrade their professional skills. In the past, most of them had only junior college or bachelor degrees, while few of them got master degrees. But now, in the Academy of Agricultural Sciences of Henan Province, dozens of colleagues have doctoral degrees and over 150 have master degrees. Besides, many of them have graduated from overseas prestigious universities and have politely declined their instructors' invitations to stay on, and have also abandoned comfortable living conditions to return to China. Equipped with modern testing instruments and adopting scientific methods in the superior environment, many researchers have made remarkable scientific achievements. Wang Shuguang is now feeling the pressure and is starting to improve English and update her professional knowledge in the evenings.

She moved home three times from 1978 to 2008. In 1978, Wang Shuguang was allotted a modest 12-square-meter one-story house. The

Academy was located in a suburb for the convenience of conducting agricultural research, but was only three km from the downtown area as the city was not big at that time. The house which she had been allocated was unpleasantly cool and humid, and was overrun with mice. The winter was the hardest time to live in the house. The piercing winds penetrated the house through wooden windows and door cracks, so she had to light the stove to keep warm. She had to pour boiled water on the frozen water pipe and tap it for a while in order to get running water. There was no refrigerator, so she hoarded Chinese cabbages underground. In 1993, for the first time, they moved into an old 50-square-meter building equipped with natural gas and heating system. In 2000, the whole family moved into a new 120-square-meter flat located on Wenhua Road in Zhengzhou. Their long-cherished wish had come true at last.

Wenhua Road is near Erqi Road at the south, and the Erqi (February 7) Memorial Tower stands erect on the southern end of the Erqi Road, which is also the city center. Although her family has moved into the downtown area, it's still four to five km to the city center. What impressed Wang Shuguang most about Zhengzhou were two dark open sewage gutters along the Huayuan Road. In the early 1980s, the sewage discharge system was not covered, and sewage gutters blocked the narrow streets, making them reek with unpleasant odors. Nowadays, the Huayuan Road has been integrated with Zijinshan Road, running from south to north in Zhengzhou. Towering highrises on both sides of this broad street and dense trees on the sidewalks make this area a symbol of Zhengzhou. The streets are becoming wider and the city is becoming more beautiful. Wang Shuguang and her family have more options to choose from when they go out. The bus station is very near to Wang Shuguang's home with dozens of all-directional bus routes. By the way, bus tickets are also very cheap, as it costs only one yuan for a trip across the entire city, while two yuan is the cost for riding on an air-conditioned bus. And bus conductors have been replaced by automatic coin-in-slot machines.

Whenever Wang Shuguang walks around in the city center — Erqi (February 7) Memorial Tower, she always find some historical traces. When she first arrived in Zhengzhou, there was only one two-story department store around the tower on a scale inferior to the present-day medium-sized supermarkets.

Since the reform and opening-up, commercial and industrial development has complemented each other, releasing tremendous energy. In the early 1990s, Zhengzhou was the scene of a nationally famous "commercial war." Kicking it off was the Asia Supermarket which became a pioneer in China's retail trade by introducing bombardment of ads on TV and novel promotional activities. Today, there are numerous large-scale markets, including Gome, Suning, Beijing Hualian and Kingbird, clustered around the Erqi (February 7) Memorial Tower together with commercial pedestrian walkways located in this prosperous area. Besides, retail industry tycoons like Wal-Mart, Carrefour, Metro and K-mart have also flocked into Zhengzhou. In those foreign merchants' eyes, this old land can bring them infinite fortune. What all of this competition brings to the macro-economy is economy of scale and efficiency and to the people it brings substantial benefits. The community where Wang Shuguang lives covers only five square meters and is now dotted by large-scale markets, including Dennis, Century Lianhua, Carrefour and Gome.

Thirty years ago, daily necessities made up the biggest expenditure for each family. Grain and non-staple foodstuffs were bought and supplied by using coupons issued by the government. Each person could only buy 26 *jin* (1 jin = 1/2 kg) of grain and a half *jin* of edible oil per month. Colleagues would run around passing on the news if grain stores offered two more taels of oil on festivals. At that time, bicycles, watches, sewing machines and radios, were symbols of wealth. If a girl could marry into a family possessing these four items, people would say she was very lucky to be able to marry such a rich man. Because of their low salaries, most families couldn't afford a bicycle. Even if you could afford one, you still wouldn't be able to get it unless you had enough manufacturers' coupons. Each family would only be issued two manufacturers' coupons per year, so, you had to collect these coupons for five consecutive years for a bike. In 1981, Wang Shuguang's family bought a 12-inch black-and-white television set which made her neighbors and friends jealous. Her small courtyard was always filled with people who would come over to watch a few TV programs in the evenings. In 1986, color TV sets were in vogue. Consequently, color TV sets were soon in short supply as panic buying prevailed. Even if you had money, it was still hard to buy one. As a result, her family asked one of their relatives who worked in the foreign trade department for help. Finally, they were able to buy an 18-inch Sanyo color TV set, and the family was in a state of excitement for a whole month.

This period during which there were shortages of goods has disappeared forever. Modern electric appliances have been added to Wang Shuguang's family's list of possessions, including a Plasma TV, desktop computer, notebook PC, Mp3, and air-conditioner. Kitchen and bathroom appliances have also undergone a "transformation": the early ventilator has been replaced by range hood; traditional iron and aluminum pans have been replaced by induction cookers, microwave ovens and automatic rice cookers. The "transformation" has made cooking more convenient and easier.

Not long ago, electric bicycles became the most popular vehicle in the city. People were fond of them because they are pollution-free and convenient. In recent years, as the number of private cars increases, the road's load-bearing capacity in Zhengzhou is facing a severe test, just as in other cities. Traffic jams have become the biggest problem. Seeking a solution, the city government has put into effect a traffic plan to build up Line 1 and Line 2 subways by 2013. By that time, the efficiency of public transportation will be doubled. Every family living in Zhengzhou is looking forward to that day.

The enhancement of the standard of living is reflected not only in the abundance of materials, but also in people's cultural life and sense of happiness. Before 1980s, everyone longed to get the chance to go on a business trip to Beijing, Shanghai and other large cities, because in that way people would have some free sightseeing. Today, having an outing to other destinations has already become a main pastime. Wang Shuguang and her family have been to Mount Tai, the Three Gorges on Yangtze River, Tianchi Lake in Changbai Mountain, Gulangyu Island, Huangguoshu Falls, Suzhou and Hangzhou, Guilin....

In 2007, what made Wang Shuguang proud was the fact that her husband who works in the Communist Party Committee Office of Henan Academy of Social Sciences, and her son who works as an editor in a financial publishing house, were both appraised as model workers in their respective organizations. This gave extra joy to the family during the coming Spring Festival.

The plane was arriving at Sanya Airport. Wang Shuguang expected to see great changes that had taken place in Sanya.

Creating a Brand Name "Single-Handedly"

Tan Chuanhua:
1978: Apprentice at a clinic in Yuexi District, Kaixian County, Sichuan Province
2008: Chairman of the board of Carpenter Tan Holding Ltd., chairman of the Chongqing Arts and Crafts Association, and director of the Chongqing Disabled Persons' Association

Tan Chuanhua had an accident in his teens in which he lost his right hand and a chance to continue his senior high school education. He had to stay at home, practicing his handwriting with his left hand, and teaching himself to draw. Because of his father's growing concern about his son's future, he was sent to his home village clinic to become an apprentice. Three years at the clinic were not enough to make a doctor of Tan, so his father managed to secure for him a position at the village school, vacated by some teachers who had returned to cities. In the winter of 1979, Tan wanted to marry his fiancé Yun Qiong and asked permission from her parents. Although Yun Qiong agreed, her parents made it clear to him that "You are just a community teacher, with no stable salary. What if you cannot maintain a teaching job?" Tan thought: "They are right. Without bread, how can you win love?"

During the Spring Festival in 1980, a distant uncle, who served in the army as a truck driver in Fengning County, Hebei Province, told Tan about many things he had experienced and heard during his trips to and from Beijing. A desire arose in Tan to see the world outside his village. His father agreed, while Yun Qiong kept silent, neither stopping nor supporting him. Early next morning, Tan went out, with a heavy bag of books and notebooks in his hand, and an easel on his back. His father walked a long distance with him and gave him 50 yuan before bidding farewell. After a short tour of Beijing, he began to study photography from a fellow villager in Fengning County, but this was not a successful experience. In the summer of that year, he went to see his cousin, the boss of a construction team in Miquan, Xinjiang, who hired him as a painter. This was a hard job for a disabled person with no right hand. Tan did not want to do this type of work, and decided to return to his hometown, for which his cousin gave him 50 yuan for traveling expenses. That year, the railway at the section of Tianshui was ravaged by flooding, so Tan had to transfer several times, and soon the 50 yuan fell short of his needs. Tan managed to arrive at Lanzhou, then proceeded on through Wu'an, Lintao, Wenxian, Guangyuan, Chengdu, Kunming to Gejiu. On the way home, he had to do painting to make ends meet. In the summer of 1982, Tan roamed from Gejiu to Liupanshui, a coal 'capital,' where he became bedridden after a bad cold. This situation kindled in him an even stronger homesickness. His original idea of painting the scenery on his way home had turned into a desire to get back home as soon as possible. Tan learned from TV news that severe flooding had occurred in his hometown Kaixian County, which made him even more anxious to return home.

After returning home, since the County Family Planning Commission was shorthanded, he was temporarily hired to do publicity work. When his fiancé's parents and relatives saw that, they were pleased and no longer opposed to the marriage. After he had worked in the town for two months, he married Yun Qiong, without an auspicious date chosen or any of the usual marriage gifts given, such as a bicycle, a sewing machine, a radio and a watch. The next day, Tan's father gave him 100 yuan, which could be regarded as a marriage gift.

When the publicity work was over, he went back to his village and set up a factory making prefabricated products. Although he was leading a prosperous life, he nonetheless yearned to venture out into the world. In 1988, he and his wife left for the city, carrying their bedrolls on their backs, taking their older son by the hand and holding the younger one in their arms. They made a living, with the husband doing paintings on glass and the wife nailing the frames. Their business went so badly that they could not even afford their rent. Suddenly, an idea occurred to Tan. There was no small merchandise sold in the town, so Tan used all their money to buy silk flowers and flower vases. The first batch of these commodities attracted so many boys and girls that all were sold out within three days. For three years they dealt in the silk flower business, enabling them to earn a profit of over 100,000 yuan and buy an apartment in town.

Later, a growing number of flower stores opened, and as a result Tan was hardly able to earn any money. In 1992, following his second brother, Tan took his family to Wanxian County. That day, they watched a TV program describing the story of Zhang Guoxi, who had become successful with his wood carvings. His second brother encouraged him, saying: "With a foundation in fine arts, you could also achieve success in this kind of business." Hearing this, Tan immediately gathered a few woodcarving craftspeople and set up the Three Gorges Handicrafts Factory with 30 employees. During the daytime, they came up with designs to carve; during the evenings, they would invite professors from the nearby Three Gorges Fine Arts Institute to give lectures. With great enthusiasm, everyone studied and worked hard, trying to carve products of high quality and artistic taste. When Tan selected some of his best woodcarvings to take part in the Shenzhen Expo, he found that, since he had not conducted any research in the woodcarving field or any market research, he had wasted half a year as well as his entire savings.

During the Expo, Tan managed to find time to visit Gorgeous China in the Overseas Chinese Town, which is a must-see tourist attraction in Shenzhen with miniature replicas of famous landmarks in

China. There he saw a number of exquisite works of art in a shop selling souvenirs. There he asked, "What sells well in your store?" "Bamboo walking sticks and wooden combs," a shop assistant replied. Although he had thought that the former would sell briskly, it had never occurred to him that the latter would be in great demand. On the way back to Wanxian County, it occurred to Tan that everybody needs a comb, which is a necessity for every home. Plastic combs create static, so there must be a market for wooden combs, perhaps a large one.

In 1993, with a bank loan of 200,000 yuan, Tan started his company manufacturing wooden combs at a factory building in Shuanhekou, Wanxian County. Nearly a year passed, and his combs were finally produced. Tan sent four of his best employees out to sell the combs. At dusk of the first day, one of them informed him that he had earned only two yuan. Compared to the large bank loan, the two yuan were hardly worth mentioning. However, this was his first business transaction. Even today, he still has the two yuan displayed in his factory's exhibition hall. The combs sold at the beginning had no brand name. When the brand name "Carpenter Tan" was later chosen, its name resonated well with the public and spread fast. Tan likes this name very much because it states his profession and carries a Chinese flavor.

In 1997, when Chongqing became a municipality directly under the Central Government, Tan was set to expand his business. However, an unexpected difficulty arose. With no fixed assets to be used as a mortgage, his small enterprise could not obtain a bank loan. Despite enormous efforts, it was all in vain. He became so furious that he reported his enterprise's difficult situation to a media agency in Chongqing. On August 18, 1997, a report, entitled "Carpenter Tan Seeking Funding from a Bank," was published. At that time, it was highly unusual for a private enterprise to look for bank sponsorship. Both domestic and foreign media followed up on the report, triggering a wide-ranging discussion on the relationship between banks and enterprises. Finally, Carpenter Tan Company obtained a loan of a million yuan and at the same time grew in popularity. To cope with fierce competition, Tan registered "Carpenter Tan" as a trademark. During the Spring Festival in 1998, to everybody's surprise, Tan invested a million yuan from the China Construction Bank in raising his product's popularity.

At first, Tan's combs were sold in stores. High distribution costs arising from store sales, along with bad debts from some poorly run stores or intentional delays in payment, put great pressure on suppliers' fund chains. As Carpenter Tan rose to fame, other comb manufacturers began to struggle for their own market shares, thus intensifying competition in stores. On March 7, 1998, Carpenter Tan signed a contract with its first franchise store. By early 2000, the number of such franchise stores had grown to almost 100. In the spring of 2000, Tan had few new stores joining his franchise. The franchise stores began to complain about low profits in comb sales. Their complaints focused on three aspects: 1) customers did not have many choices because Tan's combs lacked diversity and had only one style; 2) the prices for Tan's combs were very high, while Tan had not fully developed the brand's added value by also targeting at high-end market; 3) despite the combs' fine quality, the stores had unappealing decor on the business street. With few customers, the stores consequently got low returns from their investments and barely earned any money. Some stores were in a deficit position and even had to close down. As an entrepreneur, Tan was sensitive to market changes. He decided to spend the money he originally planned for a luxury office building and a villa on buying superior equipment and improving products' cultural content. Tan invited professors of fine arts from Chengdu to create a new decor for his stores, combining traditional and modern elements, and with a theme of traditional Chinese culture. He also invited the well-known Yumingyang-led Expert Group to introduce the Corporate Identity System (CIS) into his stores. Although these large expenditures took up one third of that year's profits, the total sales more than doubled. From then on, Carpenter Tan maintained its strong development momentum. Under the guidance of its core concepts — "honesty, labor and satisfaction," Carpenter Tan established over 650 franchise stores, both domestically and internationally, including the United States and other countries in East Asia and Southeast Asia, with the output value rising from 55 million yuan in 2003 to 140 million yuan in 2007.

Finally, Tan had a luxury office building in northern Chongqing and owned a villa, which was not so pretentious though. But he found it difficult to change his simple lifestyle, as plain as his brand Carpenter Tan. In 2007, Tan became a devout Christian. During weekends, he would invite his Christian friends to enjoy the service of the Salem Coffee Shop in his newly-opened Carpenter Tan's Handwork Store. Although he is wealthy, Tan never shows off, saying that he is just managing money for God. With a growing business, he emphasizes that he focuses only on big business deals, and the biggest deal is between God and him.

"Life is Changing and Weddings are Changing"

Xie Changjin | 1978: Physician in Henan Provincial Epidemic
Prevention Station
2008: Deputy director of the Wedding Service
Association of Henan

His colleagues and friends in the wedding service industry like to call him "Changjin." Thirty years ago, he simply helped his friends announce ceremonial procedures at weddings, but now, he is a professional MC. He said: "I've been an MC for so many years and it gives me a deeply satisfying feeling: Life is changing and weddings are changing. If you want to see a symbol of the times, you'd better take a look at the weddings at that period ."

In 1978: announcing the procedure in front of a portrait of Chairman Mao

According to Xie Changjin: "At that time, it wasn't called a wedding ceremony. And the procedure for the wedding ceremony was written on a half piece of red paper posted on the wall beside the portrait of Chairman Mao! There was no formal designation for the individual announcing the wedding procedure. People would usually ask me to announce." Changjin was extroverted and was never afraid of speaking in public, so he was often asked to be the "announcer."

In 1978, at the wedding of his friends Cheng Kaige and Zhang Guihua, standing solemnly in a Chinese tunic suit on the platform, Changjin announced: "Firstly, light the fireworks and the new couple then must move into position; secondly, bow three times to Chairman Mao; thirdly, bow to your parents..." "By that time, some traditional rituals had been abolished, like kowtowing to worship Heaven and Earth, kowtowing to parents and kowtowing to each other." Changjin said: "Actually, what I needed to do was just to say a few words for a few minutes."

Changjin said: "People were poor in those days and couldn't buy expensive items for the wedding. Three big items including a bicycle, sewing machine and watch as well as a radio set were called the 'three rotators and one sounder.' Basins, notebooks, pillowcases and bedsheets were also necessary items for a dowry. For any new couple at that time, it was not easy to assemble those items, because people could only buy them with special coupons."

It was said that one bridegroom was almost prevented from marrying his bride because he hadn't acquired a watch, a very important betrothal gift at that time. He and his wedding procession drove a walking tractor for over 60 km to his mother-in-law's home to receive the bride in the early morning. When they got there, his friends helped to transfer those articles which had been purchased onto the walking tractor, including the bicycle, sewing machine and radio set. At that very moment, his mother-in-law strongly forbade her daughter to step out of the door, declaring: "No watch, no wedding!" The reason was that she didn't see the wristwatch promised to her daughter as one of the betrothal gifts. Finally, the bridegroom had to borrow a "Shanghai" watch from someone in his wedding procession, whereupon his bride could go with him, and then, the wedding could proceed smoothly.

As for the wedding banquet, it was quite simple with close relatives and friends having a meal at home, and there were no candies served. During the period of the shortage of goods, it was rare to serve candies at a wedding.

In 1985: collecting coupons before marrying

In the 1980s, the TV set was added to the list of the most important betrothal gifts on the basis of "three rotators and one sounder." At that time, courting became more and more open and people gradually became more open-minded on the subject of marriage. A young man and young woman, who were colleagues, would gradually fall in love with each other after working together for some time. There were no romantic touches about their love. They were seen going to movies and strolling streets together. They always kept a respectable distance from each other, never even holding hands.

There were so many twists and turns in preparing for the wedding ceremony. Changjin can clearly remember that coupons were needed to buy anything from Chinese cabbage, bean curd, cloth, leather and even a box of matches, to having a bath at a public hathhouse (no bath facility at home), haircut, etc. When you got married, you would be supplied with some meat and sugar coupons from the government upon showing your marriage certificate. In any case, those coupons were insufficient for a proper wedding ceremony. So, Changjin and his family had to be busy borrowing coupons three months before the wedding ceremony. By the way, the available furniture hadn't kept pace with the development of time, and many new couples wanted to find carpenters who could make them a complete set of composite furniture to suit their new house which had an area of less than 20 sq m.

It was convenient and economical to build an open-air shed, very popular at that time, to entertain relatives and friends at the wedding banquet. Then, this led to the emergence of freelance chefs, who prepared wedding banquets. People usually found an empty and broad place to build a kitchen range and set up the iron cauldron, and shaded the place with a borrowed felt blanket. Thus, an open-air kitchen was built up, so that it was just like having a picnic.... Seen from a distance, the whole wedding spot just looked like a military camp.

According to Changjin's memory, the wedding ceremonies

around 1985 were the dullest. The wedding dresses had not yet been improved: the groom wore a Chinese tunic suit or a Western-style suit, and the bride wore a red cotton-padded jacket with a spotted pattern or a woolen overcoat if it happened to be the winter, with no obvious changes; however, the amount of the cash gift from each guest had gone up from two yuan to five yuan. According to Changjin, "In my opinion, the only characteristic of the wedding ceremony at that time was the feast." In those days, people measured whether a wedding was decent or not by the abundance of the wedding feast; for instance, whether there were chicken, duck, fish, meat, sea cucumber, sleeve fish, dried slices of tender bamboo shoots, etc. served or not.

1990: hankering after pomp and extravagance

The period of the 1990s was a time of change, when China had made much progress in its economic reform and young people made preparations for their marriage on their way to success. In this period, people began to hanker after the pomp and extravagance of the wedding ceremony; for instance, dining in star-class hotels, inviting an MC, and renting luxurious floats, etc. What bustling scenes these were!

In 1991, Changjin was invited to one of his friends' wedding, and at this point he had abandoned the title of "announcer" which had followed him for over 10 years, and was given a new designation: "MC." Besides, a Panasonic M7 video camera even overshadowed the bride and the groom, becoming the most eye-catching piece of equipment at the wedding. What made Changjin proud was the new congratulatory speech: "A graceful and beautiful couple is standing on the solemn stage; how sweet and happy the bride and the groom are! Hills and water are smiling and people are jubilant. What beautiful hills and water and what a beautiful couple they are! Birds are singing for you and flowers are blooming for you. Relatives and friends are blessing you and congratulating you and guests are applauding." This congratulatory speech, filled with cliches, has become Changjin's stock-in-trade over the last couple of years.

Changjin still remembers an episode which happened at a wedding. When the groom received his bride, all the relatives from the bride's family mimed scenarios from those movies and TV programs from Hong Kong and Taiwan prevailing on the mainland in the early 1990s, purposely barring the way until the groom sent out the gift money.

However, the wedding itself was not the focus; it was the betrothal furniture and appliances loaded on the vehicles in the wedding procession which were of most concern. The time of "three rotators and one sounder" had passed, and, indicative of the advent of the well-off society, the three important articles were now a color TV set, a refrigerator and a washing machine, which would be placed in the newly-painted house.

Taking a picture in wedding attire is one of the most important ways for the bride to record her beauty. At that time, in addition to those state-owned ones, photo studios from Hong Kong and Taiwan also entered the mainland market one by one. Photo studios could offer all-in-one service including hairstyle design and make-up. The bride looked very radiant and energetic after the stylist had created a new look for her. The effects of a kind of very blurred lens which caused a certain lack of image fidelity was quite popular with most people.

Era of the 21st century: customized wedding marches on stage

Nowadays, as living conditions have improved, people have begun to have more intellectual pursuits and to pay more and more attention to the wedding ceremony so that it leaves a lasting memory. In

2001, at the early stage of the wedding service market in Zhengzhou, Changjin, investing jointly with his friends, registered the first wedding service company in the city. This could be considered to be the starting point for the specialization of the wedding service.

Changjin also remembered the first wedding prop in Zhengzhou — an iron heart-shaped candleholder. Later, he improved this prop so that it was a double-heart-shaped one and he applied for a patent for it.

In 2002, his daughter's spectacular Chinese-style wedding caused a furor in the city. The groom wore a "Zhuangyuan Gown (ceremonial robe for someone who came first in the top imperial examination in feudal days)" and rode a big sturdy horse to receive his bride. And the bride, in phoenix coronet and embroidered tasseled cape, was delivered in a bridal sedan chair. The whole wedding was performed by ten MCs and all the local media struggled to report the event.

From 2005 on, the wedding service market has consistently been extremely hot. Those who were born after 1980 have entered a period when marriages are very extravagant. With hot pursuits to individuality, couples make plans to turn their wedding into a particular entertainment performance. They have more strict standards for their wedding ceremonies and the qualifications of the MC. Changjin says that when his company gets the new couples' orders, they are always required to provide a competent MC. The MCs have to do a lot work before mounting the stage. They need to get a comprehensive understanding of the composition of the bride and groom's families, cultural backgrounds and some anecdotes related to the couple's jobs, and to plan out various activities according to the couple's interests and hobbies.

Changjin showed to visitors a wedding scheme on which numerous wedding props were listed, including a heart-shaped festooned gate, moon-shaped candleholder, semicircular champagne tower, bubble machine, four-colored light, spotlight, laser light, cold light firework, projector, etc. And what was written on the scheme was that at the end of the month, a cartoon wedding would be held. The episode was extracted from the *Sleeping Beauty* and the protagonists were the bride and the groom. So, in this case teamwork will take the place of the single MC to meet the specific requirements of the couple. It's the market that leads to the emergence of a series of new careers such as wedding planner, superintendent, lighting engineer, DJ, etc.

Resuming Her Old Career as a Presenter

| Yang Lan | 1978: Pupil |
| | 2008: Member of the National Committee of Chinese People's Political Consultative Committee (CP-PCC), senior media personage and famous presenter of multiple TV programs |

In Yang Lan's mind, a harmonious and tranquil family is the only safe harbor upon which she can rely. When she was born, her father was working in Albania as an aid expert. When Yang Lan was a baby, her mother brought her from Beijing to Shanghai to her grandmother to be raised. In 1978, she was sent back to Beijing and to her parents whom she had missed every night and day. This family happiness which came so late is the most precious thing for Yang Lan.

As a result of the poor condition of her mother's health, by the time she went to a senior high school, she had practically undertaken all the housework, such as sweeping the floor, washing bed sheets, doing grocery shopping, cooking, buying gas in containers, etc. Making decisions for the whole family and arranging for the family's daily life at the age of no more than 16 have fostered Yang Lan's independent personality. During the period of middle school, Yang Lan was a diligent student and often achieved the first rank in her studies, not only in her school but in the whole district. At that time, one of the ways for a boy to express his affection for a girl was to pass her a slip of paper. Other girls often received this kind of paper, however, what Yang Lan received read as follows: "What's the answer to X question?" After she graduated from university and took part in her class reunion with her old classmates, Yan Lan asked her male classmates why they had always kept their distance from her during the middle school period. They replied: "You studied hard and went home immediately after class was over. So, we thought it was hopeless to pursue you..."

In 1990, an accidental opportunity allowed Yang Lan to become a TV presenter. When she was about to graduate from Beijing Foreign Studies Institute, the general director of *Zhengda Variety Show*, a CCTV program still under preparation, was casting about for a presenter in Beijing's universities and colleges. Thanks to her counselor's good opinion of her, she was recommended for the interview. She had never considered being a presenter, so she hadn't previously thought much about this career. Actually, there was no career of "presenter" until this designation appeared at the end of a TV program *Observation and Thinking* on CCTV on July 12, 1980. At that time, the style of the presenter was closer to that of an announcer. And the two most important criteria for selecting a presenter were appearance and voice. Yang Lan stood out among interviewees for her natural and fresh style, composed and decent manner and fabulous talents. Despite this, it took another six tries before she passed the screen test and even then was put on the waiting list. And then, Yan Lan spoke frankly with the director: "I dare not say that I'm very beautiful, but I do say I have all the necessary qualities and a brain. If I am given this opportunity, I hope to become an intelligent presenter." It was her frankness and persistence that finally helped Yang Lan to become the presenter of the *Zhengda Variety Show*.

The *Zhengda Variety Show* was the earliest variety show on TV for pure entertainment in China. Yan Lan looked like a young student at her debut. She was so nervous, sweating all over, and her hands were cold, as she later recalled. TV programs were few and not so interesting then, and except for news programs, there were only a few programs for daily life, a few imported dubbed films and homemade TV drama series. The *Zhengda Variety Show* was like a breath of fresh air and it attracted hundreds of millions of viewers and obtained their recognition. Every Saturday, after watching *CCTV News*, every family would switch to CCTV-2 to watch the *Zhengda Variety Show*. When those promenading outside heard the title music of the *Zhengda Variety Show* but were unable to go home to watch it on time, they would find a small store or go to their friends' homes nearby to watch it. Children would mime Yang Lan's gestures on the program, such as tilting their heads and stretching out their hands, saying: "If you want to know this wonderful world, please keep your eyes on our program." Adults would compete with each other to see who answered more questions on Monday morning. At that time, the audience was very tolerant, and never fussed about Yang Lan's slips of the tongue, just regarding them with smiles. However, Yang Lan's parents would record every session of the program with a recorder and would note down all her mistakes. This support and encouragement from her parents gave Yang Lan an infinite feeling of warmth.

The *Zhengda Variety Show* should be regarded as the ladder of life that helped Yang Lan move to maturity as well as being the huge turning point in her fortunes. In 1994, Yang Lan, having been a presenter for only four years, received the first Golden Mike Award, the highest honor for presenters in China. Yang Lan was the most excellent TV presenter in the eyes of most of the audience, especially in the eyes of the young. Her four-year-long career as a CCTV presenter had broadened her outlook and helped her set a clear goal for her future development: to be a real professional in mass media.

As one of the most popular CCTV presenters on television, Yang Lan presided over almost all important evening entertainment programs. At one time, six presenters were rehearsing for the Spring Festival Gala, but, after several rehearsals, the directing team suddenly decided to dismiss one of the presenters. On that very day, the dismissed presenter, without being told about the decision, had come into the tiring-room in evening dress in high spirits; at that moment, the make-up man told her she was not on the list. Then, she had no choice but to go away disappointed. Yang Lan, who was sitting nearby, abruptly experienced a sense of crisis. And from then on, she began to broaden her outlook and the scope of her knowledge, and warned herself to never again wallow in flowers and applauses.

In January 1994, after completing the recording of the 200th special program of the *Zhengda Variety Show*, Yang Lan boarded a flight to the USA, starting a whole new life. In her second year in the USA, Yang Lan married again. At the same time, she was studying at the Columbia University Graduate School. She seldom mentioned her first marriage. All she would say about her first marriage was that it should be put down to her rashness at a young age. During her time as a student in America, in cooperation with the Shanghai Oriental TV Station, she produced *Yang Lan Horizon*. The *Yang Lan Horizon* was a special program about American politics, economy, society and culture. And this was the first time that Yang Lan was able to comment on world affairs in her own words. She served simultaneously as the planner, producer, copywriter as well as presenter. The 40-part *Yang Lan Horizon* were circulated to 52 provincial and municipal TV stations across China. Yang Lan successfully transformed herself from an entertainment presenter to an inter-disciplinary professional in mass media.

In January 1998, *Yang Lan Studio* was formally broadcast on the newly-established Phoenix Chinese Channel in Hong Kong. In this "studio," Yang Lan worked as not only the presenter but also the manager. She selected topics on her own and budgeted very carefully for all the expenditures of her program. At first, she couldn't adapt well, because on the mainland, what you need to consider is how to make the program well, and as for the costs and profits, there are special departments taking charge of that. However, it was a really hard-won experience for Yang Lan to handle all of this by herself. Later on, this kind of operational mode was widely adopted by most TV stations on the mainland. In the following two years, Yang Lan interviewed over 120 celebrities. Those heavyweights offered her rare resources of personal relations for her future development. Exchanges with guests of different backgrounds from different industries broadened the scope of her expertise to a great extent. Two years later, she made an important transition. Now, what she lacked for marching into the business circles was only capital. Anyway, her husband was a master-hand in the operation of capital.

In October 1999, Yang Lan resigned from Phoenix Satellite Television. After remaining in the background for four months, she suddenly announced that she would purchase the Leung Kee Holdings Ltd. and rename it Sun Television Cybernetworks Holdings Ltd., successfully realizing the back-door listing. On the night of August 7, 2000, her satellite TV channel featuring history, culture and biographies — Sun Satellite Television Channel — was being broadcast in Hong Kong. Yang Lan always stood in front of the Sun Satellite Television Channel, facing the audience in high spirits. Familiar with the mass media resources, Yang Lan brought many advantages to the Sun Satellite Television Channel. However, a short while later after Yang Lan had established her business, the global economy slumped. She had to rack her brains to produce good programs as well as calculate how to ensure the company's profits. In the third year, the Sun Media could hardly collect investments. Yang Lan still chose to hang on despite all the kind warnings suggesting that she transform the business. Her husband supported her all the while, and again they invested a large amount of money. But, within three years, the company had suffered an accumulated deficit of over 200 million Hong Kong dollars. In 2005, they donated their jointly-shared holdings, accounting for 51 percent in the Sun Media Group, to society. They also resigned all positions they had held, including the Chairman of the Investment Board of Directors of the Sun Media Group. Yang Lan was very apologetic to those investors and admitted frankly: "I'm not a good businessperson after all." Consequently, Yang Lan decided to go back to the production of cultural TV programs in which she had considerable expertise.

She resumed her role as presenter at the peak of the rapid development of China's TV industry. Satellite channels, cable channels and digital channels, and even foreign satellite television channels were competing fiercely in channel specialization, uniqueness and perfection. Intelligence is more important than a pretty face for a presenter. The audience now requires more in the way of a presenter's extensive and practical knowledge and sense of humor than in his or her appearance. In 2005, Yang Lan began to host *Her Village*, a large-scale talk show specific to China's urban female audience, on the Hunan Satellite TV. At the beginning, *Her Village* was only targeted at those urban women aged 25 to 38. She never expected that the program would become so popular with both old and young, just as the *Zhengda Variety Show* had been. The content of the program just covers their experience of life and work, and existing problems. Every Saturday evening, some old fans of the *Zhengda Variety Show* will watch Yang Lan's new program , listening to this petite but talented woman talking about problems concerning themselves with the guests. To most others, the image of that yet to mature girl, tilting her head and stretching out her hands, saying: "If you want to know this wonderful world, please keep your eyes on our program," would flash back to their minds when they accidentally watched reports about Yang Lan.

Teaching – The Long March of a Disabled Teacher

Wang Shengying

1978: Primary school teacher in Xiping Village, Hengshui Town, Linzhou City, Henan Province
2008: Primary school teacher in Xiejiaping Village, Hengshui Town, Linzhou City, Henan Province

A student of Wang Shengying did the calculation for her: she walks at least 12 kilometers every day to commute between home and school and pick up her students. That means that she, with a paralyzed right leg, has covered 108,000 kilometers over the past 30 years, two and a half times the length of equator.

Mrs. Wang was born into a poor farmer's family. In order to pay for her schooling, her father risked his life working at a quarry, and her mother stayed up late at nights spinning. When she graduated from high school in 1974, she heard that the primary school in the Dongping Village was looking for a teacher. She applied, and was hired. She has since never left her teaching job.

Wang started as a "locally paid teacher," the term referring to those extra teachers not on the regular state payroll. She made five yuan a month at first, a sum too meagre to support herself, not to mention to help her parents.

She was later shifted to the primary school in Xiejiaping Village, a cluster of neighborhoods segregated by either mountains or rivers. Wang's home was 30 minutes on foot from the school, with a river in between. Because of her disabled leg, taking a tumble was a routine occurrence in her daily commute.

In the summer of 1978, the region was hit by a heavy and prolonged rainstorm, which imposed severe threats to the dilapidated building of Wang's school. She then told her pupils: "Stay at home until the rain is over. I will give you lessons one by one in your homes." In the following 20 days she made her way to all the six neighborhoods of the village to teach her students one by one, stumbling and falling again and again on the way. When she showed up at her pupil Wang Minshu's door, soaked and muddy, the girl's parents were moved to tears.

Wang's salary rose to 20 yuan by the mid 1980s, but her financial condition didn't see any improvement. Her family sank into debt after the funerals of her grandparents. At the time construction workers from Linzhou City were famous throughout the nation for their skills, and could easily find well-paid jobs out of the region. In 1985 Wang Shengying's husband contracted a construction project in northeastern China. He tried to persuade his wife to go with him: "Go to the Northeast with me. I will watch the workers, and you can handle the accounts. You won't have to work that hard, but you can make more money. It is much better than being a teacher here." Wang understood her husband's love and concern for her, but did not agree: "What will happen to my students if I leave? I became their teacher because my predecessor resigned. How can I do the same thing to them?" Seeing his wife's determination, Wang's husband left for his project alone.

After the other teacher left, Wang Shengying became the only teacher for the four grades in her school. The classrooms had been transformed from an abandoned warehouse with a leaking roof, peeling walls and dented floors. Students wrote on planks erected by earth piers, which would collapse from time to time.

While Wang Shengying was struggling with all her difficulties, her husband returned home out of concern for Wang and their children. He had since worked as a volunteer at her school, doing all the repair work. With their own money, the couple soon patched the roof.

Li Zenghua was a boy in the village whose legs were both paralyzed. When he asked his father to send him to school, the man worried that the boy might add to the existing burdens on Wang Shengying, who herself was crippled. On hearing this, Wang remembered her childhood and her pledge to never give up on any child as long as she was a teacher. "Entrust him to me," she told Li's father. "Though disabled, he has the same right as other children to receive education."

By that time Wang's school had moved to the second floor of Wang's home. Every day she carried Li Zenghua up and down the stairs on her back, and attended to his needs. As the boy grew bigger and heavier, the climb became more and more difficult. One day, both the teacher and student fell down the steps. Wang's husband happened to come back from the field at this time. He helped Wang Shengying up, and carried the boy upstairs. From that day on, he took over the task of carrying the boy up and down to and from the classroom. In this way, Li Zenghua finished his four years at Wang's school.

How the school moved to Wang's home is a long story. When Wang got to her school in the morning after an overnight rainstorm in the spring of 1993, she found the warehouse, which accommodated her school, was flattened and in shambles. There was no house in the village big enough to take in 50 children. And it was too dangerous for the children to walk along the paths frequented by wolves to attend classes in neighboring villages. Seeing no other option, Wang Shengying set up outdoor classes by the ruins. But soon the children were being viciously attacked by lice from the sheep in a nearby field. After a brief stay in several locations, Wang decided to take the students to her home. The four classes of 52 children filled both of the two rooms of her house, and spilled out into the yard. To find a final solution to this dilemma, Wang considered building another floor on top of her house to shelter her students.

The plan seemed unrealistic for a family deep in debt, which had been incurred by the husband's illness in 1989 and the construction of their current house in 1991. With a monthly pay of 40 yuan, Wang had no idea where she could get the required sum to double her home. But the couple knew that the students desperately needed a place to have classes, and they felt obliged to find one for them by whatever means they could. They sold everything valuable at home — wheat, beans and corn, and managed to buy bricks and tiles on credit and borrowed some timber from a relative. But there was still a big gap in their budget. The couple saw no choice but to borrow from siblings, six of Wang Shengying's and seven of her husband's. Their debt accumulated to 30,000 yuan, and was not paid back until 2000.

To save the cost of labor, they did all the building work themselves, laying bricks and mixing cement and sand. Their hands and knees were all covered with blisters. Out of concern for the safety of the children, Wang Shengying asked her husband to line the stairways with iron rails. When the five rooms on the second floor were finally completed, and the 52 students happily moved in, the couple broke down from the long period of overwork. The wife lost 17 kilograms during that period.

In an effort to boost the level of education in its rural areas, in 2001, China introduced the policy of providing free textbooks, eliminating tuition fees and providing subsidies for lodging among rural students in primary and junior middle schools. In 2003, with an allocation of more than 50,000 yuan, the local government established the Xiejiaping School by merging several minor schools in the region, and built nine classrooms. In addition, it was assigned four more teachers. Wang Shengying and her students were eventually able to move out of her home.

Between 1993 and 2004, more than 200 students graduated from Wang Shengying's home school. Fifteen of them later went to college, and six became village cadres.

In the fall semester of 2004, the Xiejiaping School was required to offer an English course, but none of its teachers knew enough to teach

English. Wang, approaching 50, brushed up her English and eventually became the only English teacher in her school, pending the arrival of a more qualified English teacher. Meanwhile she helped to set up a remote-education room in the school in the hope that "these lovely rural children can have access to the same educational resources as their peers in cities."

Wang Shengying's salary has grown to 1,200 yuan a month so far. She lives in the school during the week, and sees her family only on the weekends. Both her daughter and son are in college, and they and their father continue to be as supportive of Wang's work as they have ever been.

Since 1978

Zhang Yimou	1978: Worker in the Eighth State-run Cotton Textile Mill of Xianyang, Shaanxi
	2008: Director of the Guangxi Film Studio; General Director of the Opening and Closing Ceremonies of the Beijing 2008 Olympic Games and the Beijing 2008 Paralympic Games

Zhang Yimou's 30 years as a film maker is itself like an amazing legendary film: he's won nearly a hundred awards of various types; no other director in Asia exceeds this number. In the film circles of the world, not many people have been able to achieve success as both cinematographer and director in the way that he has. In addition, his films have often been huge box-office successes. With a high level of artistic and commercial value, his films have conquered the markets all over China and the world. He has hence earned a big name throughout the world.

In May 1978, after the period of social disorder, the Beijing Film Academy once again began recruiting students on a regular basis. Zhang Yimou, who had won the first prize in the National Photography Award, was quite happy to hear this news. In his journey to Shenyang on business trip, he brought his works to the Beijing Film Academy and hoped to enter this school to study. The teachers of the Department of Cinematography liked his 60 pictures on scenes in Mount Huashan. They said: "Your potential in cinematography is really great, but as you are 6 years older than the standard starting age, we are sorry that the school cannot accept you." In despair, Zhang nonetheless wrote an appeal to Huang Zhen, then the Minister of Culture. From his portfolio of photographic works attached to the appeal, Huang judged that he was talented, and instructed the admission office of the academy as follows: "Although this man cannot participate in the exam because of his age, he is talented in cinematography. I hope the academy can make an exception to allow him to study for a two-year period in the Department of Cinematography. When he graduates, he can be assigned to the Chinese Newsreel and Documentary Film Studio to engage in news or still picture photography."

In July 1982, Zhang graduated from the academy and waited for his work assignment. "Obey the assignment of the Party; a true man should start his career anywhere (he is assigned to)." This was the slogan hanging in the assembly hall when students graduated. Obeying the assignment was the standard of moral behavior for the graduates of that time. Though the best graduates could stay in Beijing to work on special assignments, there were only very few who could get permission to stay in Beijing if they came from outside Beijing. The policy at that time required all graduates to return to their hometowns to participate in the relevant work. If their hometowns or cities had no film studio, the assignment office would assign them to the film studio in the nearest city. The responsibility of the assignment office was to balance the desires of the students with the needs of the employers. For those from outside Beijing, the practical choice was to find a better film studio outside Beijing.

The Guangxi Film Studio had, at that time, just been established. Because this studio was located in the undeveloped hinterland, no student was willing to go there. The studio especially sent a manager to negotiate with the academy. They hoped the academy would be able to send a group of graduates to Guangxi. The manager promised that since there was a dearth of directors in Guangxi, when young graduates went there, they would quickly be able to make their own films. If these youths were to go to studios in Beijing, Shanghai or Changchun, they would have to work their way up through positions like assistant and associate director. It would take at least ten years before they would have a chance to make a film independently. Heads of the academy and the department all agreed to send the best talents to Guangxi. So, Zhang Yimou was assigned to the Guangxi Film Studio as a cinematographer.

Besides him, Zhang Junzhao, Xiao Feng and He Quan were also assigned to the studio. Because of the dearth of professionals, they got the chance to make a film independently by the end of 1982, after they joined one film as director's assistants. The four young people were overjoyed. They went to the writer's department and got a play script entitled *One and Eight*. They read this play and decided to shoot a film based on this play. In the studio, great expectations were placed on these young directors. They were permitted to modify the play and make a so-called "poetic film." Although they disapproved of the ending of the film, the studio leaders nevertheless let them do it their own way. The leaders hoped this film would be creative. The four young men decided to make a film defying the conventions of academic Chinese films, but expressing personality and diversity. They spent eight months preparing, including observing and learning from real life, collecting materials and selecting actors and actresses. They travelled around North and Northwest China, expending a great deal of effort in developing every single feature of the film. They treated this film like their child. When the film was first released to the public, they sat at the back of the projection room. When they saw through the window that the audience were standing up and applauding excitedly, they knew they had succeeded.

One and Eight was praised as the first work of the fifth generation of Chinese filmmakers. Naturally, Zhang Yimou became their representative. This group of filmmakers had experienced the social disorders in their youth. They were once sent to rural areas and experienced the hard times of the Cultural Revolution. When China adopted the policy of reform and opening-up, they were trained at a professional academy. They graduated from the Beijing Film Academy in the 1980s, and went on to make films with great aspirations and passion. They are extremely sensitive to new ideas and new artistic techniques, trying to express thoughts from new angles through film. They have striven to be unorthodox in their choice of subjects, narrative methods, character's description, lens employment methods and picture management. They have a strong desire to explore the history of the national culture and the structure of the national mentality. Today, this "fifth generation" is still the mainstream of Chinese film, determining the development of Chinese film.

In films like *Yellow Earth* and *The Big Parade*, Zhang, as the director of photography, was praised for the power and symbolism of his pictures. Later, he began directing films himself. His works such as *Raise the Red Lantern*, *Story of Qiuju* and *Keep Cool* have won awards at important film festivals. He continues to explore film style and language, and to introduce innovations. His films lead the way in Chinese cinema.

In the late 1980s and early 1990s, China's film market changed abruptly. Those films trying to seek new artistic values were too highbrow to attract viewers, and middle-aged and young directors diverged in the artistic pursuits.

However there were still some films which managed to achieve both artistic exploration and audience favor. *Red Sorghum* is such a film. This film, marking the directorial debut of Zhang Yimou, depicts the simple, wild life of Chinese peasants in the 1920s and 1930s, and their bloody fights of resistance against invaders. Upon its release, the film garnered both domestic and international acclaim. In 1988, it won the Golden Rooster Award and the Hundred Flower Award in China, as well as eight international awards including the coveted Golden Bear at the 1988 Berlin International Film Festival.

He was accorded the designation "Youth Expert with Outstanding Contributions" by the Ministry of Personnel of the PRC in 1993.

At the turn of the new century, Chinese cinema got bogged down once again. Nevertheless, Zhang Yimou remains optimistic and confident. He believes that with massive investment, in the future the success of Chinese cinema at the box office will definitely surpass that of foreign films. The Chinese nation has its own history and cultural traditions. It absorbs, comprehends and assimilates imported foreign cultures. Chinese people won't be interested in foreign films forever. When the feelings of excitement and curiosity toward foreign films abate, Chinese films with local and traditional flavors will surpass foreign films in popularity. In the late 1990s, Zhang Yimou and other "fifth generation" Chinese filmmakers produced a series of films with a Chinese flavor, winning international acclaim and awards. These films showed foreigners the real China — a charming, though still seemingly mysterious China.

The Montreal World Film Festival conferred the Special Grand Prix of the Americas on Zhang Yimou in 1995 for his exceptional contribution to the cinematographic arts. In 1996, he was again selected as one of the world's ten directors with top achievements in the 20th century.

Alongside the success of Chinese cinema, film critics are also very active in China. The pens of critics seemed to be the first to be liberated in the period of reform and opening-up. Some critical words leave people gasping with wonder. Zhang Yimou feels indifferent to these words and just smiles. Unconventional films and pungent criticisms are both symbols of civilization and democracy in China.

Hero, produced in 2002, was the first kung fu epic film by Zhang Yimou. It became a huge international hit, earning 177 million US dollars at the box office worldwide. It won many international awards, and was praised as a "besonders wertvoll" film by the Filmbewertungsstelle Wiesbaden. Later, his works such as *House of Flying Daggers* and *Curse of the Golden Flower* won him further international acclaim. Beginning with Zhang Yimou, the "fifth generation" of Chinese filmmakers proved through their works that Chinese films can also be part of the international market and win accolades.

Besides films, Zhang Yimou has also participated in many other types of endeavors. He once directed drama, ballet, a large-scale live performance in natural surroundings, bidding films for the Beijing 2008 Olympic Games and the Shanghai World Expo, promotion of the Beijing Olympic Games emblem and torch, and advertisements of Toyota Vios. He even assumed the role of art director of an online game... He believes that everybody has inexhaustible potential and that everyone can continuously expand their abilities through challenging themselves to the limit. This is the kind of tenacity and tension displayed by the filmmakers of his generation over these last 30 years.

The year 2008 has been a busy year for Zhang Yimou. As the general director of the opening and closing ceremonies of the Beijing 2008 Olympic Games, he and his team are undertaking a sacred task.

The grand opening ceremony of the Beijing Olympic Games held on August 8, 2008, in the Bird's Nest was a wonder beyond all imagination. Master-minded by Zhang Yimou, it unfolded a scroll of Chinese history and culture in a unique and splendid way. People will long remember the pageantry as well as the wisdom and hard work of Zhang Yimou and his team.

Reborn after the Earthquake

| Zhou Yucun | 1978: Casual bricklayer of a builder team in Fengrun County, Tangshan |
| | 2008: Retiree of the Fengrun Construction and Development Corporation, Tangshan |

On the night of July 28, 1976, Zhou Yucun, a casual worker of a builder team, fell asleep in Tangshan. His wife Wang Shuqin, who stayed in their hometown to handle the farming work, lay on her bed, trying in vain to get her two daughters to sleep. It was very hot, and even the chickens and rabbits in the yard were irritable and restless. Wang Shuqin sat up, and drank a cup of cool well water. When she came back to bed, she suddenly felt a puff of chill. She saw flashes of light outside her window, and was puzzled: Why is there no thunder after the lightning? Why has this lightning lasted so long? Several seconds later, the flashes of light faded away. A sound that Wang had never heard before seemed to spread from under her pillow. It sounded like a cow roaring at the bottom of a well. Wang Shuqin felt a thrill of fear for the first time in her life. Although more than 30 years have passed, this sound still echoes in her ears today, as though it has just happened.

Zhou Yucun, who was in the city, realized that it was an earthquake. All of Tangshan was engulfed in fog, lime, dust, coal dust, smoke and other toxic materials which are released when a city collapses. He, like others, ran about aimlessly in horror. Aftershocks followed the quake, and kept striking the city. His only wish was to survive the disaster. In this earthquake, measuring 7.8 on the Richter scale with an approximate energy equivalent to the explosive force of 400 atomic bombs on Hiroshima, Tangshan, an industrial city with a population of over one million, was decimated. The railway station, theaters, hospitals and all other buildings were demolished...

Large numbers of laborers were needed to rebuild the city. Since the earthquake had caused hundreds of thousands of deaths, Zhou Yucun was hired as a permanent worker of the builder team. He stayed in the city and joined in the work of rebuilding. Builder teams, with the help of army troops who came to conduct rescue and relief operations, picked stones, bricks and wood from debris and built many simple-structured temporary houses. These houses were built on the base of a stone wall about one meter high and the upper parts of the houses were mainly wood, grass and mud. This kind of construction is safe and easy to build.

These houses were an unusual sight after the Tangshan Earthquake. The people of Tangshan seemed to be back to a more primitive period, living in simple wooden houses. Huts of all shapes overflowed in the debris of the city. These houses might not have been beautiful, but they provided shelter for the Tangshan residents during the aftershocks and in the following difficult years. These "simple houses" in the city didn't last long. They were only in use for two to three years, but some of them in the suburbs lasted as long as ten years.

While the city was totally in ruins after the quake, the rural areas were only further impoverished. At that time, Zhou's family had four children. Since their mother had to go to work during the day, the third and fourth children were left in a wooden footbath with diapers fastened at their waists. At noon, the elder sisters would make some soup for them to drink. Supper was also insufficient, just a sweet potato and clear porridge for everyone. Only during the Spring Festival, would the production brigade allot some meat to the residents. In 1977, every member of the Zhou family was given 150g of meat and in 1978, the amount rose to 250g. Later, when land was contracted to each farm household with output quotas, Zhou's family was given

10 *mu* of paddy field and two *mu* of dry land. Their lives changed for the better, at least in terms of not being hungry. When the children grew older, they studied during the day, and helped hoe up weeds, cut firewood, and harvest rice. The life in rural areas was very hard, and Zhou and his wife hoped their children would be able to go to and live in the city through study. The family was poor and the primary school in the village only had four grades, but they borrowed some money to enable their children to study in the senior primary school in the neighboring village.

By 1986, the 10th anniversary of the Tangshan Earthquake, 230,000 Tangshan families had moved to new houses. With this, the era of the "simple houses" ended and a modernized city appeared. In that year, Zhou began taking charge of the builder team. He used discarded material from the simple houses to build a simple home. His wife, mother, and children moved into this home. The family was reunited. His little son was surprised by this kind of house that he had never seen. Holding him with one arm, Zhou pointed at the city center and said: "The office of the secretary of the CPC county committee is the same as this." His son couldn't believe this, so Zhou led his family into the city. On the road, the family went into a photo studio and took a picture of the whole family. Later, when his children went away to study or to get married, he would give them this picture. In his mind, a happy life for his family began from that time.

In 1993, he was given an apartment about 106 square meters in size by the corporation. Because Tangshan was in reconstruction, he earned much as a project manager. His wife, Wang Shuqin, got a job in the bus station of Fengrun County, and his oldest daughter also found a job after graduating from a technical secondary school. The family now had enough food and clothes. In the next year, his second daughter passed the entrance exam to the local university. His third daughter passed the entrance exam to the Sichuan Meteorological College in 1996, and his son entered the China University of Political Science and Law in 1999. Zhou was very happy that his children had not let him down and worked with extra motivation to earn more money.

The elderly couple bought a 138-square-meter house in the Fengrun District, which comprised the former Fengrun County and the Tangshan Northern New District. The house was not only larger than the old one, but it was also in a better neighborhood. The new high-rise apartment buildings and straight wide roads, with green lawns and blue sky, were a wonderful place to live. The Hope Plaza, the People's Square and the New Life Square were not far away from the house, and the couple was able to do exercises there. In the adjacent parks like Huanxianghe Park, Baiqu Garden, Garden of Serenity and Shimen Garden, they could take a walk. Central heating and gas also made life more comfortable. On fine days, the couple would go to the Nanhu Park in the south city for a walk. This area used to be a depression as a result of coal excavation. There are plenty of coal resources under the ground in Tangshan, and over a hundred years of exploitation had left large scars on the face of the area. The odor of garbage and other discarded items was very strong. When this area was being developed a few years back, Zhou Yucun seriously considered buying housing there, but he was still worried that nobody would buy houses there, so reluctantly he gave up. Since 1997, the government had begun leveling the land and planting trees there. This changed the view. Eventually it became the Nanhu Park and was awarded the Dubai International Award for Best Practices to Improve the Living Environment.

Zhou has finally retired and receives a monthly retirement pension of 1,380 yuan. With this and the support of his four children, he and his wife lead a comfortable life. But Zhou, who cannot bear to be idle, has opened a breeding farm in partnership with some friends. The income hasn't been bad. The couple also makes some money from their previous two houses. As part of the plan for improving Beijing's air quality in preparations for the Olympics, the Beijing Shougang Company moved its steel production to Caofeidian in Tangshan. Since then, Caofeidian has become the most popular industrial park in north China. In 2007, it was found out that Tangshan had one billion tons of oil deposits. Tangshan is on the brink of yet another great development. With abundant experience in real estate development, Zhou made a fortune from reselling houses. In 2006, he discarded the microbus that he had been using for six years and bought himself an "Accord" Honda from Beijing. Even "Accord" is regarded as an economy car in Tangshan nowadays as 78 Mercedes-Benz vehicles were added in Tangshan in 2006 alone.

The old couple didn't interfere with their children's marriages, just requiring that the mate have good moral qualities. Their first child married in 1995. This son-in-law had grown up in a rural area, and the wedding was held in the countryside. The husband's family wasn't rich, so they only gave old Zhou 5,000 yuan as a betrothal gift. Some of Zhou's friends thought it was too little, but old Zhou felt that if his daughter liked this man, the marriage was a good one despite the amount. Their second daughter married in 1999. This son-in-law also came from a rural area, but the family was richer. His family gave old Zhou 10,000 yuan as a betrothal gift. The wedding was held in the city. Two Mercedes-Benz and several Audi cars were spotted in the wedding procession and the wedding hall was set in the just opened prestigious Jingtang Hotel. This made old Zhou feel honored. Their third daughter and son-in-law married in Chengdu without a wedding and went to Sanya in Hainan for their honeymoon. Although this third son-in-law didn't give old Zhou any betrothal gift, nor did he hold a grand ceremony, old Zhou was still quite happy, as he thought this type of marriage was really fashionable. His only son has worked in a government agency in Beijing for five years after graduation. He is diligent and capable, so his leaders think highly of him. Old Zhou often tells his son, "You should know your own capabilities. Don't think too much about positions and don't be motivated by sheer vanity, or you will fail." The only thing old Zhou is worrying about now is the fact that his son still doesn't have a girlfriend. His son has said that he is seeking a girlfriend based on his father's requirements, according no importance to appearance, work or money, but only trying to find a girl with good moral qualities.

Turn Wood into Gold

Zhang Guoxi | 1978: Director of the Jiangxi Yujiang Carving
Craft Plant
2008: Chairman of the board of Guoxi Group,
member of the CPPCC National Committee
and vice-chairman of the Jiangxi Society
for Promotion of the Guangcai Program

Zhang Guoxi is not the richest entrepreneur, but he is the first in China to have a minor planet named after him. In 1993, proposed by the Purple Mountain Observatory and approved by the Committee on Small Body Nomenclature of the International Astronomical Union, the minor planet, designation number 3028, was named after Zhang Guoxi. Before him, the only entrepreneur in the world who had won this honor was Doctor Hammer. Zhang Guoxi was nominated not because he was the first multi-millionaire on the Chinese mainland, but because of the contributions that he and his enterprise have made for the common good during these past many years.

His poor childhood left him especially eager to make contributions to the public welfare. When he was only two years old, his mother passed away. When he had just reached middle school, the Cultural Revolution ended his studies. In 1967, Zhang Guoxi, only 15 years old at the time, became an apprentice at the carpentry workshop of the Agricultural Tool Making and Repairing Cooperative in Dengjiabu Town, Yujiang County. Five years later, the workshop was separated from the cooperative. Zhang Guoxi assumed the position of director. As the head of the new factory, he was given nothing from the cooperative except three pedicarts of lumber and several work sheds. He had to assume responsibility for the lives of 21 workers and their families, as well as a debt of 24,000 yuan inherited from the cooperative. When pay day came, the factory had no money at all. Zhang Guoxi sold his house, which had been passed down to him by his grandfather, for 1,400 yuan. Zhang Guoxi gave all the money to the factory as capital.

But this small amount of money wasn't even enough. Influenced by the educated youths sent to Yujiang from Shanghai, the idea of going to Shanghai to find some business intrigued him. With 200 yuan and three workers, he braved the trip to Shanghai. At night, they slept under the eaves of a department store to save money. Hearing that the Shanghai Arts & Crafts Import & Export Company was purchasing set cases made of camphorwood, he went to the company and inquired about making some such cases for them. Hearing that they had come from Yujiang County, a place made known nationally by Chairman Mao Zedong's poem, the company's manager immediately ordered 50 sets from them. This was the beginning of Zhang Guoxi's career in craftwork .

Zhang Guoxi's second trip to Shanghai was in the fall of 1979. In the sample exhibition hall of the Shanghai Arts & Crafts Import & Export Company, he took an interest in Buddha's niches for export to Japan, which were much more expensive than the set cases. The niches were small in size, but complicated in structure. They had several thousands of parts of different sizes and shapes. If any part went wrong in terms of size or shape, it became impossible to assemble the niche. The requirements in craft techniques were very high, so many factories didn't dare to undertake this type of business. But Zhang Guoxi reasoned that this product cost little but its value was high. This would be a business turning wood into gold. Leading a technical group himself, Zhang Guoxi led this factory to finally produce this kind of niche and then sold them to Japan. In the early 1980s, his personal capital had reached 30 million US dollars. He was the first multimillionaire in China.

To participate in the international market, Zhang Guoxi maintained strict standards for product quality. He once broke several hundred yuan's worth of beam just because the beard of the Chinese dragon on the left of the beam was 2 cm shorter than the one on the right. Another day, a Japanese merchant made a claim for compensation because he said a niche had fallen apart during the journey to Japan. Zhang Guoxi led this merchant into the warehouse and had him choose any of the available products for the purpose of a test. A worker held the packed product chosen by the merchant two meters above the ground and dropped it onto the concrete floor twice. He then opened the wooden packing and had the merchant examine it. Although the wooden packing had broken, the niche inside was still in good shape. By maintaining good quality, Zhang Guoxi's enterprise beat his rivals from Taiwan and South Korea, despite their advanced techniques and level of experience. The Buddha's niche business was greatly expanded through his efforts. He later set up the Jiangxi Craft Industry Co. Ltd. to produce lacquer. He began selling lacquer products instead of semi-finished wood craft products. Not only did the added value of the wood carvings increase significantly, but the competitiveness of these products in the international market was also significantly enhanced. Meanwhile, through establishing the Association for Sales of Buddha's Niches in Japan, he pulled those minor enterprises in Japan together to sell his products.

Starting in 1990, Japan's economy experienced a recession. Especially during the 1997 Asian Financial Crisis, Japan's economy was at its lowest point since World War II. Given such conditions, the Buddhist products market also cooled. Many enterprises producing or selling Buddhist products closed down. But Zhang Guoxi had a different view on this stagnancy. His analysis was that this period of stagnation in Japan wouldn't last too long. As long as Buddhism was still the belief of the Japanese, Buddhist products could be sold in Japan. Therefore while the market remained in the doldrums, Zhang Guoxi increased production capacity instead of withdrawing from the market. On the foundation of one factory, he invested another 40 million yuan to set up two other factories. As the economy in Japan gradually revived two years later, the need for Buddhist products also resurfaced. The sale of his products increased and about 60 percent of Japan's Buddha niche and spares market was taken up by the Guoxi Group. His enterprises began to lead the market.

The diversification of the Guoxi Group was initiated in Hainan. As the largest special economic zone in China, Hainan became a separate province in 1988. The Hainan Island became a focus for all the people of China. People rushed to the island. They were of different ages, in different professions, with different cultural backgrounds and with various wishes and expectations. From morning to night, people crowded onto every ship across the Qiongzhou Strait that separates the island from the mainland, and even more people were waiting in the Hai'an Port. Everyone was saying that one can "earn money in Hainan; make a fortune in real estate." Zhang Guoxi himself was one of the several thousand deputies voting for the proposal to set up the Hainan Special Economic Zone at a session of the Seventh National People's Congress. After the NPC session, he immediately decided to invest in real estate in Hainan. He crossed the Qiongzhou Strait together with a stream of other people and stepped onto this "golden land" with the spirit of a pioneer. He looked about everywhere in Haikou, and determined to invest 200 million yuan to build a grand hotel in Sanya, a land with wonderful sights.

Later, Zhang invested another 100 million yuan in Hainan to develop tourism and real estate projects. However, as the Hainan fever gradually cooled down, its economy was experiencing a sluggish market. Moreover, in 1993, the whole country was facing a difficult time from an economic standpoint. The Retail Price Index increased 10 percent over the previous year, and the exchange rate of Renminbi to US dollar dropped 45 percent. The policy makers judged that the economy was overheated with high inflation. In this year, the country began winding down and rectifying the real estate market. The real estate bubble burst after experiencing sharp expansion. Most real estate dealers owed money to banks. These dealers paid their debts with real estate, but kept the money they had earned. The tropical and subtropical scenery of Hainan and especially Sanya is best represented by the beach facing the South China Sea. Huge rocks on the beach bear characters which mean in English the "End of the Sky and the Rim of the Sea." Now people mockingly listed unfinished buildings as a new tourist attraction. After 1994, the Hainan economy slumped further. At every Spring Festival, a large number of people left Hainan and didn't return. Many famous corporations and figures disappeared. The craze of investment cooled down, and the tourist market was also stagnant. In 1996, Zhang Guoxi had to suspend construction of his project. After 2000, once the tourist market in China was heating up again, Zhang Guoxi immediately went to Sanya himself to restart the hotel project. A mere half a year later, a four-star hotel was completed and opened to tourists.

In addition to wood carving products, real estate and tourism, Zhang Guoxi also invested in auto spare parts, the liquor industry, micro electric machinery and the stock market. It is his operational mode to gradually expand the group with appropriate diversification while sticking to the main line.

Zhang Guoxi never regards his fortune as something important. He wouldn't feel ashamed to be poor, nor think highly of himself for owning a billion dollars. His attitude toward money is, you can't bring it with you from before birth, and you can't take it with you after death. To use money where it is most needed is a charitable act. Zhang Guoxi allots the profits according to the principle of "paying taxes and fees first, then leaving enough for the development of the enterprise, making the employees rich and making contributions to society." His enterprises have been active in donations to various charities. In the early 1980s, when the profits of his enterprises reached a million yuan, he donated a building for scientific education at the Yujiang First High School. Later, when his enterprises continued to expand, he donated money to build the county TV tower, a welfare home and the 100-meter-long Guoxi Bridge. He also actively contributed money to charities like flood relief and the Candlelight Project. Zhang Guoxi and his enterprises have donated over 40 million yuan to society.

Despite becoming rich early in life, Zhang Guoxi can't forget his poor childhood. When he married, he couldn't support the family at all as his business was just taking shape. His two daughters were brought up on the income of his pharmacist wife. Though her income was small, his wife still managed to save 10 yuan every month for him as pocket money. At that time, Zhang Guoxi often worked late until three to four in the morning. But whether it was cold winter or hot summer, his wife would get up and boil hot water or cook some food for him. Now he has become as rich as a Forbes billionaire, enjoying banquets almost every day. But he still always recalls the food his wife cooked for him in those hard days. He admonishes his entrepreneur friends: "Don't talk about divorce after you get rich, not even you were 100 times richer." He still maintains his sincere and honest character.

Living Beside a Venue of the Olympic Games

Jia Guohua | 1978: Returning to Beijing city as an educated
youth after being sent to Beijing East Subur-
ban Farm during the Cultural Revolution
2008: Member of the Chinese Communist Party
Working Committee and head of the Organization Department, Datun Sub-District Office in Chaoyang District, Beijing

As usual, Jia Guohua comes home at eight in the evening. After greeting his wife, he turns on the hi-fi system and listens to a few old songs sung by modern singers. He has found himself much busier in 2008 than he was before. Beijing Municipal Government and Chaoyang District Government have designated the Datun Sub-District Office where he works as "a window during the Olympic Games." He and his colleagues are honored to assume this responsibility, and are working with double care. His wife is very supportive. As the principal of a primary school, she is busy at work, but she offers the car to her husband though his office is closer to home, and shoulders almost all of the housework. According to Jia Guohua, "I have worked for 30 years, but I've never been so tired."

In 1974, Jia Guohua graduated from middle school and, being among those educated city youth sent to the countryside for reeducation during the Cultural Revolution, he went to Beijing East Suburban Farm. He stayed there for four years, mixing feed, doing farm work, driving a tractor, and working as a statistician. Compared with other young people sent to locations outside of Beijing, he was lucky to be staying in a suburb of Beijing. For this reason, he was content with his life on the farm.

In the latter half of 1978, a national working meeting on the educated youth sent to the countryside was held, after which a series of new policies were released and a large number of these young people would go back to urban areas. By that time, Jia Guohua had done farm work for four years. At the end of that year, leaders of the farm informed these young people of the good news that they would be ending their farm life and returning to urban areas and waiting for a formal job assignment. However, Jia Guohua was not so happy. He had been assigned to this farm after middle school and could do nothing but farm work. When he returned home near Anzhen, his parents were very happy to see their grown-up son. To welcome him, his mother cooked more dishes than usual and they enjoyed a family reunion.

Soon, he was assigned to Beijing No.1 Housing Building Company. Although he was only a temporary worker, he felt good about having a job. He mixed mud and moved bricks; tired as he was, he felt fulfilled. After returning to Beijing, that period of his life was unforgettable for him.

He worked as a temporary laborer for a short period and then went home to convalesce because he had been experiencing hypertension. Later, he passed the entrance exam at Beijing Metallurgical Plant, which was located in Datun Township. At that time, there was only a narrow tractor path (the present Beiyuan Road) leading to the outside, with vast farmlands lying on either side. On the day he reported for duty, longing to begin his new job, Jia Guohua got off a bus and walked along the path to the plant. As he had once worked in a building company, the plant leadership assigned him to the capital construction team. The head of the team realized that he was good at mathematics and made him a storekeeper instead of a builder.

Several years later, Jia Guohua decided to enroll in college as he felt that his level of knowledge was inadequate. While he was preparing for the exam, the plant was transferred to the Capital Iron and Steel Company, which had some quotas for "ownership by the people (meaning full status as employees on regular payroll)" and the plant leaders encouraged him to take the exam. Jia Guohua ranked first among more than 80 examinees, and was transferred to a workshop. In 1991, the company headquarters selected some people from among its subordinates and Jia was among them. Since the company headquarters was far away from his home, Jia was reluctant to go, but nonetheless accepted the assignment. Every morning, he went to the subway station near his home by bicycle, then took the subway to the station near the Capital Iron and Steel Company and then walked there. He received a 10-yuan commutation ticket from the company each month. However, he had to spend three hours on the way to work and home every day, and was thus unable to take care of his family. For this reason, he wanted to return to the plant as soon as possible.

In 1993, the company headquarters finally allowed him to return to the plant and he became a staff member of the labor union. He liked his work, which included writing posters and organizing activities for workers. In 1995 before Datun Sub-District Office was established, he was temporarily transferred to the township government to do preparatory work.

Owing to his satisfactory work during several months there, Jia Guohua was retained by the township government and was allocated a home along Ring Road Four, near the National Aquatics Center. It was his first allocated home. After marriage, he and his wife had lived in a seven-square-meter room in his parents' home. After his daughter was one month old, they moved into his father-in-law's 16-square-meter house in a small hutong near Andingmen, and lived there for five years. Later, several old classrooms at the school where his wife worked were subdivided to temporarily solve the housing problem of some teachers who had no housing. And thus they lived in half a classroom for seven years. When dilapidated buildings in the Anzhen area were demolished, his mother got a new apartment near Beishatan, bigger than her old home. Jia Guohua and his family moved again to live with his mother. This time, Datun Township Government allocated him a three-room one-storey house with a toilet and kitchen, covering an area of 100 square meters. Jia Guohua built a small garden and made a parking space in front of the house, ready to live here for several years.

His daughter studied at a secondary school for dance. Jia Guohua loves his daughter very much. On weekends, he would always take a taxi van to bring her home and send her back to school on Monday mornings. Later, taxis vans were upgraded to Charade vehicles, which were more expensive, and consequently Jia Guohua decided that he would like to buy a car. In 2000, his wife suggested that they buy a Santana. A short time later, their salaries rose and Jia Guohua discussed the matter with his wife and decided to save more money and buy a Citroen ZX the following year. At the end of 2002, Jia Guohua received his year-end bonus and immediately bought a Jetta GT. When his wife came home from work, he handed her an empty envelope. "Where is the money?" she asked. Jia Guohua pointed to the new car in the parking space outside the window, and said, "I bought the car with it." The couple smiled. Now, his daughter has graduated from the Dancing College of the Central University of Nationalities and works at the China Opera and Dance-Drama Theatre. Jia Guohua has bought a 240,000-yuan Beijing Hyundai Tucson and had planned to give his five-year-old Jetta GT to his daughter. But his daughter doesn't want it and would like to buy a car of her choice.

With the expansion of Beijing, Datun Township became part of downtown Beijing, and many high-grade residential quarters and social organizations were built here. Old farmland gradually gave way to wide roads and towering buildings. In 2001 Beijing won its bid to host the 2008 Olympic Games, and as planned, half of the Olympic Village and Olympic Green was to be located in Datun Township. The nation and the world immediately focused its attention on Datun Township. It may take several hundred years to complete the process of urbanization-modernization-internationalization in other places, but it took only seven years in Datun. The township government had more work to do. Jia Guohua was appointed deputy-director of the Resettlement Office and his first task was moving and resettling his own house where an Olympic venue was to be set up. Although he encountered complex problems in the process of removing and resettling residents, local farmers and residents were cooperative because they were enthusiastic about the coming Olympic Games and enjoyed favorable resettlement policies. In 2005, Datun Township Government became Datun Sub-District Office, the first in Beijing in over 50 years.

Jia Guohua also saw the substance of his work change from rural construction to urban management. He worked simultaneously in three departments: city management, comprehensive control and Party work, and saw the remarkable improvement in city infrastructure and residents' qualities and felt the urgent demand of the people for qualified grassroots civil servants. In 2004, he started to acquire computer knowledge from the keyboard, and to learn 300 English sentences for civil servants from ABC. In the same year, he received his Bachelor of Law diploma. The biggest change for him was that he upgraded his concept of democratic management. At the end of 2006, Jia Guohua, member of the Party Working Committee and head of the Organization Department, took part in directing the community committee elections at the expiration of the previous office term.

At a community near the major Olympic venues, the candidate for community head recommended by Datun Sub-District Office lost in the democratic election. At first Datun Sub-District Office members were surprised at the result, but they finally approved the election result. After careful and delicate communication with the new community head and staff, the community work went off normally and over a year later, some aspects of the community work even earned a name for the Sub-District Office.

Now Jia Guohua is at home and his wife has finished cooking and has placed the dishes on the table. There are eight dishes, though small in size but balanced. His wife serves him rice and pours a cup of medicated wine for him while she asks, "What did you do today?" Jia Guohua pauses for a few seconds before answering: "Trivial things for residents, but all related to the Olympic Games. And I feel great!" The couple smiles at each other. This is his life, peaceful and content.

From a Champion to a Billionaire

Li Ning	1978: Athlete of the Gymnastics Team of Guangxi Zhuang Autonomous Region
	2008: Chairman of the board of Li Ning Sporting Goods Co., Ltd., vice-chairman of Chinese Gymnastics Association, and vice-chairman of China Sporting Goods Federation

In 1999, Li Ning was elected one of the Best Athletes of the Century by worldwide online voting, together with another 25 sport stars including Muhammad Ali, Pelé Eterno and Michael Jordan.

A famous scholar once observed that most of the elite writers in China were from ethnic minorities. For instance, Lao She was ethnic Manchu while Shen Congwen was ethnic Miao. Actually, some famous Chinese athletes are also from ethnic minorities, such as Li Ning who is ethnic Zhuang. Li Ning began formal gymnastics training in his native Guangxi at the age of seven.

Li Ning's success was closely connected with the Olympic Games. In 1978, China's legitimate seat in the International Olympic Committee (IOC) was restored, and as a result, China's status in the international sporting community rose suddenly and athletes had more opportunities to participate in world competitions. In that year, Li Ning had trained for eight years, but had not yet gone to the national team. In 1979, Chinese gymnasts attended the 20th World Gymnastics Championships after 17 years of interruption. The 16-year-old Li Ning longed to participate in the following Championships.

In 1980, Li Ning had his dream fulfilled and was chosen by the national team. Honor then came to him in succession. In 1982 at the Sixth World Cup Gymnastics, he won six out of the seven gold medals for men, making a new record in the world history of gymnastics, and thus was acclaimed as the "Prince of Gymnastics." In 1984, he won three gold, two silver and one bronze medals at the 23rd Los Angeles

Olympic Games, becoming the athlete who won the most prizes at the games. In his 18-year athletic career, from 1970 when he entered a sports school to 1988 when he retired, he won 106 gold medals including 92 in Chinese competitions and 14 in world competitions. In 1987, he was elected a member of the IOC Athletes Commission, the only member from Asia at that time.

Before 1988, Li Ning was at the zenith of his fame and got all the glory; he was regarded as a hero by the Chinese people and was always in the spotlight surrounded by flowers and applause. The Seoul Olympic Games was deemed a Waterloo for China after returning to the Games, since many Chinese gold-medal favorites finisihed poorly. The people had staked all their hopes on Li Ning. With a huge amount of mental pressure on him, Li Ning suffered a defeat during the last competition, falling down from the rings. At that time, people were not as open-minded and as tolerant as they are today and could not accept such failure as they had staked all their hopes on Li Ning, a world champion. When he returned to the Capital Airport in Beijing, Li Ning deliberately avoided the spotlight and flowers and chose another exit. At the other end of the passage was Li Jingwei, general manager of Guangdong Sanshui Jianlibao Group, who had come all the way from Guangdong and was holding flowers to welcome Li Ning.

After the Seoul Olympic Games, Li Ning announced his retirement. For some time after his retirement, he did not know what to do. He had two choices, just as many other retired athletes have: to be a coach or a government official. Guangxi Sports Commission had invited him to take the post of vice president; a foreign country wanted to invite him to be the coach of its national team; and some entertainment company had also extended him an invitation. Li Ning himself wanted to set up a gymnastics school in South China to train more talented athletes for the country. When Li Jingwei learned of his idea, he told Li Ning that he would need economic support to develop sports so that it could be a stable and long-lasting program. Although at that time, public opinion did not agree with the idea that a world champion should engage in business after retirement, Li Ning became special assistant to Li Jingwei who had urged him to do so. Several months later, Li Ning got the idea of starting a sportswear factory. In 1990, Jianlibao Sportswear Company was set up, with capital from a Singaporean company and Jianlibao Group, and Li Ning being the general manager. The sportswear was branded "Li Ning."

In the same year, the Tenth Asian Games was held in Beijing. Shrewd Li Ning realized that this was the best opportunity he had to promote his brand. Although the style of "Li Ning" sportswear was not finalized and the quality was not the best among similar products, he hoped that all the torch bearers could appear around China wearing "Li Ning" sportswear so that all the Chinese people would know the "Li Ning" brand and get familiar with it. He, a former world champion, went to the Finance Department of the Asian Games, but was informed that the department had offered a price of US$3 million for the hosting of the torch relay and a Japanese financial group and a South Korean group intended to buy out the rights. His career as an athlete had made him a person who never gave up easily. He spared no effort in meeting with the organizing committee of the torch relay several times, marking his change from an athlete to a brand franchiser. The "Li Ning" brand and Li Ning's skills in public relations underwent a big test, but the result explained everything. Finally, the Organizing Committee of the Tenth Asian Games decided to make Jianlibao Group the sole sponsor of the torch relay with three million yuan, and to designate "Li Ning" sportswear as the clothing of torch bearers, honorees of the Chinese teams and all the journalists, forging a path for Chinese sporting goods brands. When "Li Ning" sportswear made its debut, almost all sports lovers found it dazzling and the brand was widely accepted.

Li Ning succeeded the first time. In 1991, with an investment of 16 million yuan by Jianlibao, Guangdong Li Ning Sporting Goods Company was formally set up to specialize in managing sportswear and "Li Ning" brand sports shoes. At the end of 1992, three branch companies were set up in Beijing and Guangdong to specialize in producing and managing sportswear, leisure wear and sports shoes respectively. Since getting into business, Li Ning has persisted in the goal of "basing business on sports and contributing to sports." The "Li Ning" brand has always supported the development of world sports, especially Chinese sports. Chinese athletes wearing sportswear and shoes branded "Li Ning" could be found at the Barcelona 1992 Olympic Games, Atlanta 1996 Olympic Games and Paralympic Games, Sydney 2000 Olympic Games, and Athens 2004 Olympic Games... Li Ning Sporting Goods Company was also the sponsor of some national teams, such as the Chinese Gymnastics Team, Shooting Team, Diving Team and Weightlifting Team.

Since 1991, the turnover of the "Li Ning" brand had been increasing by 100 percent every year. However, there was a hidden problem. Li Ning Sporting Goods Company was a full-invested subsidiary company of Jianlibao Group, whose controlling stockholder was Sanshui County People's Government of Guangdong Province, state-owned assets. Li Ning expected his company to develop into a modern corporation while Li Jingwei supported him to set up his own company. From 1993, Li Ning kept on buying out the 16 million-yuan stocks held by Jianlibao. In September 1994, Li Ning Sports Industry Company was formally established, and initially joined the capital market as a holding parent company. At the end of 1995, Lin Ning Group was set up and Li Ning himself became chairman of the board and general-manager. At the beginning of 1996, Li Ning moved the head-quarters from Guangdong to Beijing in a clean severance of the last ties with Jianlibao. In that year, the sales revenue reached 670 million yuan, a historical jump.

As competition grew sharper, the "Li Ning" brand met with crisis. Although brand awareness was still high, "Li Ning" products were of low quality and old styles and the company management was in chaos. In 1997, the sales stopped growing; and in 1999, the sales revenue dropped from US$76 million in 1997 to US$60 million, ringing an alarm bell for Li Ning. On the one hand, he fired some family members from important posts and employed more professional managers to manage the group, so as to turn the family business into a modern corporation. He invited experts in capital operation to serve as independent directors. In 1999, he hired several companies specializing in different fields to help him complete an overall strategy; SAP of Germany, Maida of Taiwan, BICI Beijing Investment Consultants Inc., Dentsu Inc. of Japan and PricewaterhouseCoopers were Li Ning Group's partners. By doing so, Li Ning aimed to strengthen the international competitiveness of the group and put the promotion of the brand around the world on the agenda.

On the other hand, Li Ning had the idea of getting the group listed. In December 1998, to pack the best capital of the group and have it listed, Shanghai Li Ning Sporting Goods Holdings Ltd. was set up in Pudong New Area, which later became the major part of the listed company. One year later, however, the company, with its management and ownership structure not sufficiently defined, could not meet the requirements for a listed company, and thus Li Ning did not get the company listed in Hong Kong. In January 2003, Tetrad under Government of Singapore Investment Corporation (Ventures) Pte. Ltd. and CDH Fund under China International Capital Corporation entered Li Ning Group, holding 19.9 and 4.6 percent of the stocks respectively. Thus, the ownership structure finally was transformed.

A series of reform measures saw good results. With the turnover growing and capital entering the company, Li Ning Company was formally listed at the main board of the stock market in Hong Kong on June 28, 2004. On that day, the volume reached 25,300,000 shares, making it a top performer on that day.

An important reason why Li Ning chose to get the company listed in Hong Kong was that he hoped to accelerate the internationalization process of the "Li Ning" brand. In August 1999, the "Li Ning" brand took part in the World Sporting Goods Exhibition in Munich, Germany for the first time as a representative of the Chinese sporting goods industry. This shows that the Li Ning Company began its entry into the European market with a sports brand. In 2000, the "Li Ning" brand had franchises in nine European countries including Spain, Greece and France. In the same year, Li Ning defeated Adidas and won the bid to sponsor the French Gymnastics Team. In 2008, the "Li Ning" brand became the official apparel for Chinese table-tennis, diving, shooting and gymnastics delegations, Spanish men's and women's basketball delegations to the 2008 Beijing Olympic Games. Li Ning has set another goal: to occupy 20 percent of the world market in 2018 and to make the "Li Ning" brand an international sports brand.

Perseverance Has Brought What I Have Today

Liu Yonghao

1978: Lecturer of the Sichuan Institute of Machinery Industrial Management

2008: President of New Hope Group, member of the Standing Committee of the Chinese People's Political Consultative Conference, and the largest shareholder in China Minsheng Banking Corp., Ltd.

Liu Yonghao, the fourth son in his family, had the biggest wish to eat twice-cooked pork once a week and Mapo tofu every other day. These popular local dishes represented the ideal life pursued by Liu Yonghao in 1966. Two years later, he went to live and work as part of a production brigade at the Gujia Village, Xinjin County in the suburb of Chengdu, capital of Sichuan Province. The village had neither electric power nor water supply, not to mention a complete highway. His daily pay was measured in work points worth only 0.14 yuan. At that time, country life was hard and the farmers were poor, and it was even worse in a major agricultural province like Sichuan. Before graduating from Sichuan Agricultural University, Liu Yonghao's third eldest brother told his mother that he wanted to be a farmer after graduation and return to his home village to do breeding. Looking at her third eldest son, who had been brought up by another family due to the poverty of his own family, his mother replied in total bewilderment, "You have been a farmer for over 10 years. Don't you know about the hardships of the countryside?" To please his mother, the filial son became a technician at the agricultural bureau of Xinjin County, but he did not give up his idea of breeding.

From the first to the seventh day of the Spring Festival in 1980, Liu Yonghao's second eldest brother Liu Yongxing set up a stall on the street for repairing TV and radio sets for the purpose of earning enough money to buy some meat for his four-year-old son. As it turned out, in a few days Liu Yongxing had made 300 yuan, equivalent to his 10 months' salary. The four Liu brothers were bursting with joy over their amazing good fortune. Since repairing wireless receiving sets could earn them so much money, they decided to start an electronics factory. Manufacturing electronic products was an easy matter for the Liu brothers, given that the eldest, Liu Yongyan was a university graduate in computer science; the second, Liu Yongxing, was a repairman thoroughly acquainted with home electrical appliances; and the fourth, Liu Yonghao, was a teacher at the Institute of Machinery Industrial Management. China's first home-made hi-fi was soon produced and they named it Xinyi, meaning novelty. Liu Yonghao took their hi-fi to the countryside to seek cooperation with a production brigade in order to set up a factory, in which they would offer techniques and management while this production brigade would provide capital, and each partner would own half of the factory. Then their application was submitted to the commune. Unfortunately, the Liu brothers' plan was aborted because of a simple comment from the Commune Party secretary to the effect that collective enterprises must not cooperate with private individuals or follow the "capitalist path."

Their passion for enterprise had been kindled despite their failure in starting a hi-fi business. Liu Yonghao's dream of twice-cooked pork and Mapo tofu was replaced with the dream of acquiring a wealth of 10,000 yuan, which was an equivalent of a millionaire in today's terms. What business would he choose to achieve his dream? The hi-fi industry required a large amount of investment and had so many taboos and restrictions, whereas the breeding industry had a relatively lower threshold for investment and techniques and he was quite familiar with it. Therefore, Liu Yonghao and his brothers began to raise quails on their own balcony despite the scorn of their neighbors. The quantity of quails and eggs increased day by day. Liu Yonghao and his brother Liu Yongxing used their bicycles to peddle quail eggs in the streets after work. Liu Yonghao felt embarrassed about running into his students while he was selling the eggs. Nonetheless, his brothers and he ended up with a bulging purse. The Liu brothers then decided to set up a breeding farm in their home village.

As employees on state payroll, the four Liu brothers did not want to bring any troubles onto themselves. Therefore, Liu Yonghao paid a visit to the Party secretary of the local county and consulted him about whether their plan was in violation of the current policies. The open-minded Party secretary agreed to their idea, but required them to take on 10 households specialized in breading. With the county Party secretary's consent, the Liu brothers immediately applied to the bank for a loan of 1,000 yuan, but their application was rejected. They had to pool their money by selling iron scraps, wristwatches, bikes and black-and-white TV sets.

Liu Yongmei, the third son in the family, was the first to take an indefinite leave of absence from his work unit and became the manager of Yuxin Breeding Farm in his home village, followed by the resignations of the other three brothers from their enviable public posts. The breeding farm mainly hatched and raised chicks and quails and bred vegetable seeds. Owing to a lack of money, the brothers bought steel scraps from stalls and made incubators with tools rented from a factory. Liu Yonghao bought used bricks from Chengdu to build the farm's workshops and transported them to the village in a tractor. However, due to the narrow village road, the tractor could not enter the village and they had to unload the bricks two kilometers away. Liu Yonghao solicited the help of several farmers and they carried the bricks back to the village by hand and on their shoulders. By the end of 1983, the Liu brothers were excited by their inventory for that year, in which they hatched 50,000 chicks and 10,000 quails and took on 11 breeding households.

In April of the following year, a man from Ziyang County, Sichuan Province, who specialized in breeding, came to the Liu brothers and arranged a big order of 100,000 chicks. However, half of the first batch of 20,000 newly hatched chicks died from a lack of air during transportation and the rest were burnt to death in a fire at the man's home. That man went bankrupt and fled his home. His wife knelt down before Liu Yonghao and begged him to forgive her husband. The remaining 80,000 eggs would soon hatch. State unified purchase and sale had caused a shortage of feedstuff in the busy farming seasons. Moreover, their debts were due. The four brothers entertained the thought of throwing themselves into the Minjiang River or fleeing to Xinjiang. Their morale was greatly affected by their first crisis since startup. At the critical moment, Liu Yongyan, the eldest of the four, said, "We must persist." That same night, the four brothers wove baskets, with which they then carried the remaining chicks to the local farmers' market before dawn every day. To their surprise, despite their considerable loss of weight due to hard work, the four brothers managed to sell all the chicks. By 1986, the Yuxin Breeding Farm was turning out 150,000 quails annually. At first, they opened up a wholesale outlet for quail eggs in the Qingshiqiao Market, Chengdu. As their business expanded, the four brothers set up another outlet in the Dongfeng Market for farm produce, the biggest of its kind in Chengdu, supplying several hundred thousand eggs per day. They had too many orders from Chongqing, Xi'an, Xinjiang and Beijing to deal with. With impetus from the flourishing business of the Liu brothers, one third of the farming households in Xinjin County began to raise quails. Xinjin, raising 10 million quails at their peak, became a quail wholesale center in China. The Liu brothers decided to rename their breeding farm

"Hope," a word indicating a promising future.

Without any reservation they imparted their quail breeding techniques and experience to the households specializing in this field in Xinjin County. The breeding households of Xinjin collaborated in groups and bought the feedstuff and implements from the Liu brothers. Before long, they had all surpassed the four brothers in terms of hatching rate, egg laying rate and feed conversion ratio. When Liu Yonghao was in Guangzhou on business, he was astonished to see farmers in Guangzhou lined up in long queues to buy Chia Tai pellet feed. That prompted him to think about the prospects of pig feed production. The raising of pigs was a mainstay of the rural economy in Sichuan, which was a major pig-raising province in China. After careful analysis, the Liu brothers decided to shift their operation from quail breeding to feed production and they made detailed strategic plans. They purchased a tract of land of 10 *mu* (about 6,700 square meters) in Gujia Village and invested all of their previous profits into the feed production project. The Hope brand complete formula pellet feed for porkers researched and developed by their company appeared on the market in 1989, and gradually broke the previous monopoly of Chia Tai brand on the high-end feed market in China.

In 1992, the Hope Group, with the Hope Feed Company as its predecessor, was established and became China's first privately-run enterprise group approved by the State Administration for Industry and Commerce. The group's assets were equally distributed among the four brothers, each assuming his position based on interests and expertise. Liu Yonghao was President and legal representative of the Hope Group. In March 1993, Liu Yonghao was elected a member of the Eighth National Committee of the Chinese People's Political Consultative Conference and he submitted a proposal along with another 41 members for setting up a bank, which would mainly handle investments from private entrepreneurs and serve privately-run enterprises. Based on that proposal, China Minsheng Banking Co., Ltd. was founded two years later.

At the startup phase of their business, all of the four brothers strove for consensus in running their business. However, as the enterprise expanded, the emergence of disparities in their views regarding the development of their business was inevitable. The family enterprise Hope Group was restructured twice in its transition and regulation process toward a modern enterprise. On April 13, 1995, abiding by the principle of equal division of assets, the Liu brothers, in contrast with their previous ambiguous sharing of property rights, suddenly made the division perfectly clear. Liu Yongyan, the eldest of the four, set up the Continental Hope Group; Liu Yongxing, the second eldest, the East Hope Group; Liu Yongmei, the third of the four, the Huaxi Hope Group; Liu Yonghao, the youngest, the South Hope Group, which was the predecessor of today's New Hope Group. The move was subsequently appraised as the most successful example of dividing assets among brothers in the history of Chinese enterprises.

The New Hope Group is no longer a feed producing company. Assessed by the State Administration for Industry and Commerce as the largest privately-run enterprise in China, the group deals in a broad range of fields, including feed, dairy and catering, finance and investment, estate development, and the basic chemical industry. After May 1999, in one year's time, Liu Yonghao, with 9.99 percent of its shares, became the largest shareholder of the China Minsheng Banking Co., Ltd. by purchasing shares for a total of 186 million yuan. The New Hope's sustained and steady return on equity from the Minsheng Bank has become an important source of its revenues. In the early days of reform, when food and clothing were China's principal concerns, the Liu brothers were engaged in the farming industry. When the service industries began to thrive, the New Hope went into the financial industry. In 2002, the New Hope Group signed a contract with the local government of Yangshuo County, Guangxi Zhuang Autonomous Region, which entitled them to a 50-year leasehold right to the core scenic area of the Guilin landscape. In 2003 the group constructed a five-star hotel in the west hi-tech zone in Chengdu and advanced into the tourist industry. The development of New Hope is clearly indicative of China's economic pulse.

Having been for a time among those on the Forbes rich list for the mainland of China, Liu Yonghao placed first on the Hurun 2007 China Rich List. However, Liu Yonghao does not like to be called a big shot or tycoon. His austere and simple way of life seems eccentric. He dislikes Western suits and usually wears T-shirts and casual trousers, which altogether cost only a hundred yuan or so. He always buys economy class airline tickets and in more than a decade has never changed his hairstyle, during which time he has continued to patronize the same barbershop, spending several yuan on a simple natural haircut each time. Liu Yonghao has continued to work at least 12 hours a day. Study has been the main theme of his life. When he is talking with others, no matter who they are, he always takes out a notebook and records anything he thinks is useful. He still loves eating twice-cooked pork and Mapo tofu. At noon, he does his best to return home on time to enjoy the dishes cooked by his wife whenever he is in Chengdu. Otherwise, he usually has a box lunch. He is also accustomed to having lunch together with the grassroots employees in the group's dining hall. He eats fast and never leaves a single morsel in his canteen.

From Metropolis to Border Town

Xu Yulin: | 1978: Editor/reporter for the *Xinjiang Daily*
         | 2008: Retired

In a golden autumn in the 1960s, a group of young men and women who had the motherland in mind and yearned for the world left their hometowns for Xinjiang — a border region in northwest China. Among them was a slightly built man of average height wearing glasses. He was Xu Yulin, a Beijing native and an undergraduate student from the Philosophy Department of the Renmin University of China.

During the five days and nights of train travel, he spent most of the time sleeping on the floor under his seat. When he was awake, he joined the group in talking in high spirits about everything: the future, their aspirations and lives, while eating the food prepared by their parents and drinking water provided on the train. When the train was approaching Xinjiang, less and less greenery could be seen outside the window. A vast desert unfolded in front of them. It reminded Xu of a Tang dynsty poem depicting the desolateness of the northwest border region of China. He became visibly in lower spirits.

When they arrived at Urumqi, capital of Xinjiang, they were all shocked by the thickness of the snow on rooftops and ice in the streets. Xu, carrying his heavy luggage, didn't dare to move one step. Some of his classmates were bold enough to walk a few steps, but they fell down so hard on the ground that they couldn't stand up by themselves. They had to resort to tricycles.

Passing by those low and dilapidated and snow-covered houses along the anomynous streets, Xu completely lost his interest and passion. His hands and feet were numbed by piercing cold wind even though he was wearing woolen gloves and fur shoes.

Time passed quickly. He had spent two years on an Army reclamation farm in Xinjiang. The perpetually snow-topped Tianshan (Heavenly) Mountains, evergreen pine trees, vast Gobi Desert, grasslands and the crisp and cool air out in the border region had purified and nurtured Xu's mind. Work and life in those fiery and militant days slowly but steadily restored his lost passion.

After Xu left the army reclamation farm, he was transferred to the *Xinjiang Daily* to work as a journalist. Since then, he has been conducting a wide and comprehensive survey on life in the border region. Back then, journalists had a very difficult life, especially when going to remote areas such as the underprivileged minorities' villages at the south foot of the Tianshan Mountains. Inconvenient transportation, the hard lives of the villagers, the undeveloped means of production, as well as the pure rustic customs were imprinted on Xu's mind. At that time, it took about ten days to take a bus from Urumqi to Hotan in south Xinjiang. Travelers had to rush out of the bus to fight for a bed in a roadside inn after suffering a day of bumpy journey, as there were not many beds. It was common to see one room filled with many people, both old and young, men and women.

In spring, large sections of roads paved with sand and rubble partly sank as a result of thawing. The bus behaved like a boat tossed by rough waves on the sea. However, the passengers grinned and bore it.

On one occasion, Xu went with a senior editor to Minfeng, a county at the foot of the Kunlun Mountains. They traveled 2,700 km, skirting the famous "Sea of Death," Taklimkan Desert. Their task was to report on those Uyghur peasant workers who were digging an irrigation canal by hand in rocky mountains to divert water from the ancient muddy Niya River. Xu was deeply touched by those perseverant and courageous laborers. They simply used sharp chisels to dig at hard rocks little by little with dim oil-lamps lighting in a cave dozens of meters deep. They could only dig a few feet or even a few inches every day, but they never gave up. Years of effort finally saw the completion of a 10,000-meter-long tunnel for directing water to their fields. Xu was moved by their feat but also felt sympathetic and sad about the poverty and lack of modern tools there.

Xu also found his love in Xinjiang. It happened in the course of going to visit a Kirgiz grazing camp on the Chinese side of the border with the then Soviet Union. He invited Jiang Tingyan, a female secretary of the Department of Publicity of the Kizilsu Kirgiz Autonomous Prefecture to go with him. It was in August, the sky was azure blue with a few patches of clouds and the weather was cool on the Pamirs Plateau. Xu and Jiang rode horses along the picturesque canyon of the Pamirs. Everything seemed so romantic to the two young people: majestic ice-topped peaks, roaring water, wild trees and chirping birds. They fell in love with each other before his assignment was over.

As the story goes, soon afterwards they got married and had a baby girl. Life continues. Thirty years later, their baby had grown up and they had become old. Xu had left the newspaper long before. During a summer a few year ago, Xu took his daughter, who was on a summer vacation from the Communication University of China in Beijing, to south Xinjiang to retrace his footprints. The father and daughter traveled along the rim of the Taklimkan Desert in a comfortable car.

They passed through the towns of Korla, Aksu, Kucha, Artux, Kashgar and Hotan. Great changes had taken place in south Xinjiang: high-rises, green trees and fruits could be seen everywhere; highways covered almost every city in Xinjiang. Even some pockets of the unproductive Gobi Desert had been developed into vineyards and orchards, and parts of some newly-opened roads were decorated by grape trellises. His driver kept exclaiming with excitement "The traffic is so great!" Finally they arrived at Hotan, "the land of jade." They found people selling jade everywhere. In their hotel in Yutian, there was one piece of jade stone weighing 4,700 kg.

After visiting Hotan, they headed towards Xu's old friend's home in Minfeng, where he had interviewed the Uyghur laborers. This city is no longer a small desolate county town, but has become the starting point of the first road that crosses the desert. In the past, going from Minfeng to Urumqi was 2,700 km by car; but today, in addition to air links, even the cross-desert road shortens the distance between the two cities by nearly 2,000 km.

The following morning, Xu and his daughter said farewell to his friend and headed to Korla, an important oil base in south Xinjiang and the capital of the Bayingolin Mongolian Autonomous Prefecture. Along the road, they saw huge nets laid on both sides of the road for fixing sands. Further away were green belts planted by oil workers and served by drip irrigation while oil rigs dotted the land far and near.

He couldn't help thinking of the early explorers who rode camels, carrying their own water and food as they headed for the unknown wilderness at the risk of their lives.

Xu who is currently retired from his position as the director of the autonomous region's Taiwan Affairs Office, often climbs the Hong (red) Hill with his wife and has a bird's eye view of Urumqi from the hilltop. From there he sees that the Urumqi River which used to flood every year now slowly winds its way through beautiful residential and commercial areas, and nurtures thousands of acres of fields and thousands of people. In the north, the hotel which used to be the highest building in the city when Xu first came has become a government guesthouse. Now it is dwarfed by other modern buildings which have 30 to 60 floors. Even so, the people, including Xu, will not forget it as it witnessed part of the city's history.

Reminiscing about the past, Xu and his wife cherish more about their current lives and feel confident and optimistic about the future, just like their young daughter.

Postscript

The ancients said: "Thirty years is one generation; predestined kings only succeed after thirty years through a benevolent reign." This saying may serve as a suitable footnote to the glorious course of China's 30 years of reforms and opening-up. We believe that the best way to comprehend a period of history is to provide a record of the lives of various people in this period. To keep these memories alive, on the occasion of the thirtieth anniversary of the reforms and opening-up, we have compiled and published this book which truly and visually reflects the transformation of China's society over the past 30 years.

The cooperative efforts of numerous people have successfully brought this book to the world. Firstly, we should acknowledge that we owe the success of this book to the painstaking work by the team of authors, who, apart from the principal author Pan Deng, include An Ran and his A Life Devoted to the Teapot, Song Jie and her A Teacher's 30 Years, Zhu Ling and Zhou Wei and their A Bus Rider's Six Moods over a 30 Year Period, Zhao Xiaomeng and her Witnessing the Changes in Zhengzhou, Le Yan and her "Old Shanghainese" Gets Lost in Pudong, Yuan Jun and her The Long March of a Disabled Teacher, Lin Yun and her "Life is Changing and Weddings are Changing" and Fang Wen and her I Have Seen a Green China and A Veteran Construction Engineer Who Helps Change Beijing. Meanwhile, we should also express our gratitude to the people who agreed to be interviewed by the authors. It is their elaborate narrations that have helped us recapture vivid life stories over those 30 years. We are very sorry that we were unable to contact one or two intended interviewees due to time constraints, and we have organized and supplemented their life stories according to existing materials. However, we still wish to express our deepest thanks to them for adding more bright hues to the memories of those 30 years with their colorful lives.

We have surprised the world with great deeds over the past 30 years. Today, standing at a new starting point, we are still confident in our belief that miracles will continue to happen in the future.

图书在版编目（CIP）数据

30年，30人：英文/ 潘灯 等著. －北京：外文出版社，2008
ISBN 978-7-119-05439-1

I.3... II.潘... III.人物－生平事迹－中国－现代－英文
IV.K820.7

中国版本图书馆CIP数据核字（2008）第121522号

策　　划：黄友义　李振国　胡开敏
责任编辑：文　芳
翻　　译：王琴　李洋　欧阳伟萍　孙雷　曲磊
　　　　　严晶　冯鑫　周晓刚　姜晓宁
英文审定：Solange Silverberg　王宗引
图片提供：CFP　CNS　兰佩瑾　诸彪　朱玲等
印刷监制：张国祥

30年，30人
见证中国改革开放

潘灯 等著

©2008外文出版社

出版发行：
外文出版社（中国北京市西城区百万庄大街24号）
邮政编码：100037
网址：http://www.flp.com.cn
电话：008610－68320579（总编室）
　　　008610－68995852（发行部）
　　　008610－68327750（版权部）
制版：
北京维诺传媒文化有限公司
印刷：
北京外文印刷厂
开本：787mm×1092mm　　1/16　　印张：15.5
2008年第1版第1次印刷
（英）
ISBN 978－7－119－05439－1
08800（平装）
17-E-3891P